Problems and Snapshots from the World of Probability

Gunnar Blom   Lars Holst   Dennis Sandell

Problems and Snapshots from the World of Probability

With 30 Illustrations

Springer-Verlag

New York Berlin Heidelberg London Paris
Tokyo Hong Kong Barcelona Budapest

Gunnar Blom
Department of Mathematical Statistics
University of Lund
Box 118
S-221 00 Lund
Sweden

Lars Holst
Department of Mathematics
Royal Institute of Technology
S-100 44 Stockholm
Sweden

Dennis Sandell
Department of Biostatistics
  and Data Processing
Astra Draco AB
Box 34
S-221 00 Lund
Sweden

Mathematics Subject Classifications (1991): 60

Library of Congress Cataloging-in-Publication Data
Blom, Gunnar.
    Problems and snapshots from the world of probability / Gunnar
  Blom, Lars Holst, Dennis Sandell.
       p.   cm.
    Includes bibliographical references and index.
    ISBN 0-387-94161-4. — ISBN 3-540-94161-4
    1. Probabilities—Problems, exercises, etc.   I. Holst, Lars.
  II. Sandell, Dennis.   III. Title.
  QA273.25.B58   1994
  519.2—dc20   93-31828

Printed on acid-free paper.

Production managed by Henry Krell; manufacturing supervised by Vincent Scelta.
Camera ready copy provided by the authors.
Printed and bound by Edwards Brothers, Inc., Ann Arbor, MI.
Printed in the United States of America.

9 8 7 6 5 4 3 2

ISBN 0-387-94161-4 Springer-Verlag New York Berlin Heidelberg
ISBN 3-540-94161-4 Springer-Verlag Berlin Heidelberg New York

Preface

We, the authors of this book, are three ardent devotees of chance, or somewhat more precisely, of discrete probability. When we were collecting the material, we felt that one special pleasure of the field lay in its evocation of an earlier age: many of our 'probabilistic forefathers' were dexterous solvers of discrete problems. We hope that this pleasure will be transmitted to the readers.

The first problem-book of a similar kind as ours is perhaps Mosteller's well-known *Fifty Challenging Problems in Probability* (1965). Possibly, our book is the second.

The book contains 125 problems and snapshots from the world of probability. A 'problem' generally leads to a question with a definite answer. A 'snapshot' is either a picture or a bird's-eye view of some probabilistic field. The selection is, of course, highly subjective, and we have not even tried to cover all parts of the subject systematically. Limit theorems appear only seldom, for otherwise the book would have become unduly large.

We want to state emphatically that we have not written a textbook in probability, but rather a book for browsing through when occupying an easy-chair. Therefore, ideas and results are often put forth without a machinery of formulas and derivations; the conscientious readers, who want to penetrate the whole clockwork, will soon have to move to their desks and utilize appropriate tools.

The 125 problems and snapshots are presented in as many sections grouped into 17 chapters. We recommend that readers make their own very personal choice of topics. They should probably start with some of the 'Welcoming problems' in Chapter 1 and continue by taking a sample of titles from Chapters 2–16. Perhaps they will then read about rencontres, the ménage problem, occupancy, the reflection principle, birthdays, the tourist with a short memory, the irresolute spider and Markov chains with homesickness. (The last three are examples of sections that we wrote just for fun.) Finally, they may end up with one or two of the 'Farewell problems' in Chapter 17, for example, the final problem about palindromes in random sequences.

We hope that the book will acquire at least three groups of readers.

First, it is intended for 'probabilistic problemicists', people who love to solve all kinds of problems in probability, to construct challenging problems and discover truths that may inspire them or others to new research. Problemicists are found almost everywhere – among students, among teachers and researchers in probability, as well as among mathematicians.

Second, the book can be used by dedicated undergraduate students for supplementary reading after taking some regular course; much of the material is meant to extend the knowledge of discrete probability in various directions and give the student intellectual stimulation.

Third, it is our hope that the collection will be used as a reference book, since many of the results given in it are not easily accessible elsewhere. Partly for this reason, we have prepared a detailed index of subjects and names.

Two sources for the work deserve to be mentioned: the third edition from 1968 of 'Feller I', indispensable for all friends of probability, as well as the excellent 1990 book by Hald on the history of probability and statistics. Feller I is devoted to discrete probability, as is Hald's book to a large extent, in view of the predominance of discrete problems in the early days of probability.

The prerequisites for studying our book are modest. Some college courses or university courses of basic mathematics, including difference equations as well as a solid course of elementary probability are sufficient for reading Chapters 1–12 and 17. For the study of Chapters 13–16 it is valuable, though not necessary, to know something in advance about Markov chains, Poisson processes, order statistics and martingales.

At the end of many sections we have added problems for the reader, generally with answers. Some problems are meant as pure exercises, some are given as a challenge to the reader, and in some cases only vague indications of a problem area are put forward, which, we hope, will inspire interested readers to start their own constructions. Teachers may find some of the exercises, or simplified versions of them, useful for their courses.

We have done our best to reduce the error frequency in the book by several independent checks, especially of formulas and answers to exercises. We should be much obliged to the readers for information about errors and other imperfections.

When preparing the book, we have been much stimulated by the excellent journal *The American Mathematical Monthly*, especially by its problem section; some problems have been quoted from that section.

We thank Anders Hald for his valuable comments to the sections with historical content, and Jan-Eric Englund for suggesting the problem in Section 1.1 and for commenting on an early draft of the book. Furthermore, our sincere thanks are due to David Kaminsky for his detailed linguistic revision of the manuscript. Finally, we are very grateful to Martin Gilchrist of Springer-Verlag for his positive attitude to our project and for his editorial help.

Lund and Stockholm   GUNNAR BLOM
July 1993   LARS HOLST
                                                    DENNIS SANDELL

Contents

Symbols and formulas

The probability of an event A is written $P(A)$.

The probability that two events A and B both occur is written $P(A \cap B)$ or $P(AB)$.

The probability that at least one of two events A and B occurs is written $P(A \cup B)$.

The conditional probability of an event B given that the event A has occurred is written $P(B|A)$.

The complement of an event A is written A^*.

Random variable is usually written rv.

Independent identically distributed is often written iid.

The expected value of a rv X is written $E(X)$, and its variance is written $Var(X)$. The covariance of two rv's X and Y is written $Cov(X,Y)$.

By $Bin(n,p)$ is denoted a binomial distribution with probability function

$$\binom{n}{k} p^k (1-p)^{n-k}, \quad k = 0, 1, \ldots, n.$$

By $Po(m)$ is denoted a Poisson distribution with probability function

$$\frac{m^k}{k!} e^{-m}, \quad k = 0, 1, \ldots.$$

By $U(0,1)$ is denoted a uniform distribution over the interval $(0,1)$.

By $Exp(a)$ is denoted an exponential distribution with density function

$$\frac{1}{a} e^{-x/a}, \quad x > 0.$$

By $\Gamma(p,a)$ is denoted a gamma distribution with density function

$$\frac{1}{a^p \Gamma(p)} x^{p-1} e^{-x/a}, \quad x > 0.$$

By integration of the beta function $x^{a-1}(1-x)^{b-1}$ we obtain

$$\int_0^1 x^{a-1}(1-x)^{b-1} dx = B(a,b),$$

where

$$B(a, b) = \frac{\Gamma(a)\Gamma(b)}{\Gamma(a+b)}$$

and a and b are positive quantities. In the special case when a and b are integers we have

$$B(a, b) = \frac{(a-1)!(b-1)!}{(a+b-1)!}.$$

The beta distribution has density function

$$\frac{1}{B(a,b)} x^{a-1}(1-x)^{b-1}, \quad 0 < x < 1.$$

The sum of the *harmonic series* can be written

$$1 + \frac{1}{2} + \frac{1}{3} + \cdots + \frac{1}{n} = \ln n + \gamma + \frac{1}{2n} - \frac{1}{12n^2} + \frac{\theta_n}{120n^4},$$

where $0 < \theta_n < 1$; here $\gamma = 0.57721566\ldots$ is Euler's constant. For details concerning this formula, see Graham, Knuth and Patashnik (1989, p. 264). The first three terms generally give a good approximation to the sum.

Stirling's formula can be written

$$n! = n^n e^{-n} \sqrt{2\pi n} \exp\left(\frac{1}{12n} - \frac{1}{360n^3} + \frac{\theta_n}{1260n^5}\right),$$

where $0 < \theta_n < 1$. For large n we have the approximation

$$n! \approx n^n e^{-n} \sqrt{2\pi n}\, e^{1/(12n)}.$$

For details, see Graham, Knuth and Patashnik (1989, p. 467).

1
Welcoming problems

This introductory chapter contains a sample of problems and snapshots meant to convey the flavour of the book to the reader and whet his appetite for studying it.

1.1 The friendly illiterate

All the letters from a sign marked

OHIO

USA

have fallen down. A friendly illiterate puts the letters back haphazardly, four in the first row and three in the second. In doing so, he puts each letter upside down, in a vertical direction, with probability $\frac{1}{2}$ and correctly with probability $\frac{1}{2}$. (If an S is turned in this manner, it resembles a question-mark, excluding the dot.)

Find the probability that

1. the first word

2. the second word

is correct.

The problem can be solved in several different ways, one of which is shown here: Denote the two probabilities by P_1 and P_2, respectively. Note that the letters H, I and O can be set upside down without changing their meaning, but not U, S and A. We use the 'classical definition of probability', dividing favourable by possible cases.

When all seven letters are returned in random order, and put either upside down, or right-side up, there are $2^7 \cdot 7!$ possible cases, where the first factor corresponds to the turning operations.

The number of ways leading to the correct word OHIO in the first row is found in the following way. For the two O's there are $2 \cdot 2 \cdot 2$ favourable positions, including the turning operations, and 2 for each of H and I. The letters U, S and A can be put in any internal order and any of the two turning positions, which gives $2^3 \cdot 3!$ cases. Multiplying all these numbers we obtain $2^8 \cdot 3!$ favourable cases in all.

In a similar manner, we find that there are $2^4 \cdot 4!$ ways to obtain USA in the second row. Hence,

$$P_1 = \frac{2^8 \cdot 3!}{2^7 \cdot 7!} = \frac{1}{420},$$

$$P_2 = \frac{2^4 \cdot 4!}{2^7 \cdot 7!} = \frac{1}{1,680}.$$

One can see that the probability of obtaining the longer word OHIO is four times larger than that of obtaining the shorter word USA; this is, of course, due to the presence of two O's in the first word and the vertical symmetry of its letters.

If, instead, the friendly illiterate twists the letters 180 degrees with probability $\frac{1}{2}$, the answer becomes different. (An S is invariant to such a translation.)

1.2   Tourist with a short memory

A tourist wants to visit four capitals A, B, C, D. He travels first to one capital chosen at random. If he selects A, he next time chooses between B, C, D with the same probability. If he then selects B, he next time chooses between A, C, D. (His memory is so short that he forgets that he has already visited A.) Next time he chooses again between three capitals, and so on. Find the expectation of the number, N, of journeys required until the tourist has visited all four capitals.

Set

$$N = Y_0 + Y_1 + Y_2 + Y_3,$$

where Y_i is the number of journeys required for visiting one more capital when i capitals have been visited. The first two Y's are, of course, 1. Furthermore, Y_2 and Y_3 have a geometric probability function

$$p(1-p)^{k-1}$$

for $k = 1, 2, \ldots$, with $p = 2/3$ and $p = 1/3$, respectively. The mean of this distribution is $1/p$ and therefore we have $E(Y_2) = 3/2$, $E(Y_3) = 3$. Hence,

$$E(N) = 1 + 1 + \frac{3}{2} + 3 = \frac{13}{2}.$$

Note that

$$E(N) = 1 + 3\left(\frac{1}{3} + \frac{1}{2} + 1\right),$$

which suggests the expression for a general number of capitals.

1.3   The car and the goats

Suppose that you are on a television show and they show you three doors. Behind one door is a car, and behind each of the two others a goat. You are asked to choose one of the doors. You pick a door, say no. 1, which, however, is not opened. The host, who knows what is behind all three doors, opens one of the other two doors, say no. 3, and out comes a goat. (The host never opens the door which hides the car.)

He then says to you: You are allowed to switch from door no. 1 to door no. 2 if you find that advantageous.

Should you switch or not?

This question was given to the columnist Marilyn vos Savant of the American Publication *Parade Magazine*; see vos Savant (1990a, 1990b, 1991). She answered that, by switching, the chance of winning the car is doubled; the probability increases from 1/3 to 2/3. We shall prove in a moment that her answer is correct.

The answer created an intense debate among mathematicians, readers of *Parade Magazine* and others. Many debaters argued as follows: When the host has opened a door and showed you one of the goats, there are two unopened doors behind which one car and one goat are hidden. Therefore it does not matter if you change or not, since the chance of winning the car is $\frac{1}{2}$ in both cases.

We now prove that vos Savant's answer is correct. Let us examine the two alternatives just described.

Alternative 1: Do not switch

As there are three doors, the probability of winning the car is 1/3.

Alternative 2: Switch

Number the animals goat no. 1 and goat no. 2. There are three equiprobable cases for your first choice of door.

1. You pick the door with the car. By switching you lose, of course.

2. You pick the door with goat no. 1. The host opens the door with goat no. 2. By switching, you select the door with the car and win the game.

3. You pick the door with goat no. 2. Again you win the game.

Summing up, there are two favourable cases out of three. Hence the probability of winning the car is 2/3 when you switch.

The problem confirms the old 'axiom' that the land of probability theory is full of stumbling blocks. One of the purposes of this book is to help the reader to avoid them.

For interesting discussions of the TV game described above, and various generalizations, see Engel and Venetoulias (1991) and Morgan, Chaganty, Dahiya and Doviak (1991). We end the section with a generalization not mentioned by these authors.

A generalization

Person A is invited by person B to play a game with an urn containing a white balls and b black balls ($a + b \geq 3$). A is asked to choose between two strategies; he will be blindfolded before beginning the game:

1. A draws a ball at random. If it is white, he wins the game, otherwise he loses.

2. A draws a ball and throws it away. B removes a black ball. A again draws a ball. If it is white, he wins the game, otherwise he loses.

Which strategy is best for A? Clearly, if A chooses Strategy 1, he wins with probability

$$P_1 = \frac{a}{a+b}.$$

If A chooses Strategy 2 we get by conditioning on the first draw that A wins with probability

$$P_2 = \frac{a}{a+b} \cdot \frac{a-1}{a+b-2} + \frac{b}{a+b} \cdot \frac{a}{a+b-2} = \frac{a}{a+b}\left(1 + \frac{1}{a+b-2}\right).$$

As $P_2 > P_1$, Strategy 2 is the best one for A.

In the TV game, we have $a = 1, b = 2$, and so $P_1 = 1/3$ and $P_2 = 2/3$.

1.4   Patterns I

This is an elementary snapshot about patterns in random sequences. In Chapter 14 and Section 16.6 we give more general results.

Toss a balanced coin at the time-points $t = 1, 2, \dots$. Denote heads and tails by 0 and 1, respectively. We are interested in the resulting random sequence of zeros and ones. A *pattern* S is a given sequence, for example, $S = (1\,0\,1\,1)$. We shall discuss two problems concerning patterns.

Problem 1. Waiting time for a pattern

a. $S = (1\,0)$. Let N_1 be the waiting time for S. For example, if the random sequence is 0 0 1 1 1 0, then $N_1 = 6$. Find the mean waiting time $E(N_1)$.

b. $S = (1\,1)$. Find the mean waiting time $E(N_2)$ for the pattern $(1\,1)$.

The probability of obtaining $(1\,0)$, or $(1\,1)$, at two consecutive given positions in the random sequence is $1/4$. Therefore, it is tempting to believe that the mean waiting time is the same in both cases, but this is not so.

Problem 2. Waiting time between patterns

  a. Consider a random sequence beginning with $S = (1\ 0)$. Let N_3 be the waiting time until S occurs again. For example, if the sequence is 1 0 0 0 1 1 0, then N_3 is 5. Find the mean waiting time $E(N_3)$.

  b. Consider a random sequence beginning with the pattern $S = (1\ 1)$. Find the mean waiting time $E(N_4)$ until S occurs again.

In (a) the smallest possible value of the waiting time is 2, which arises if the sequence is 1 0 1 0. In (b) the smallest possible value is 1, which occurs in the case 1 1 1. (Overlapping of patterns is allowed.) Hence the distributions of N_3 and N_4 are different. However, the mean waiting times are the same, which is remarkable.

Solution of Problem 1a

Conditioning on the outcome of the first toss, we obtain $N_1 = 1 + N_1'$ if 0 occurs and $N_1 = 1 + Z$ if 1 occurs. Here N_1' has the same distribution as N_1, and Z has a geometric distribution $p(1 - p)^{k-1}$ for $k = 1, 2, \ldots$, with parameter $p = \frac{1}{2}$. Hence,

$$E(N_1) = \tfrac{1}{2}[1 + E(N_1)] + \tfrac{1}{2}[1 + E(Z)].$$

Noting that Z has mean $1/p = 2$ and solving for $E(N_1)$, we obtain the answer $E(N_1) = 4$.

Solution of Problem 1b

We use the same type of reasoning as in the previous solution, but now we have three cases: (i) the first toss results in 0, (ii) the first toss results in 1 and the second in 1, (iii) the first toss results in 1 and the second in 0. Considering each of these cases and taking averages, we obtain the relation

$$E(N_2) = \tfrac{1}{2}[1 + E(N_2)] + \tfrac{1}{4} \cdot 2 + \tfrac{1}{4}[2 + E(N_2)].$$

This equation has the solution $E(N_2) = 6$. Hence the mean waiting time for (1 1) is larger than that for (1 0).

Solution of Problem 2a

It is realized that the solution is the same as in Problem 1a. Hence we have $E(N_3) = 4$.

Solution of Problem 2b

The first two tosses result in (1 1). Depending on the result of the third, toss we have two cases. Taking the average, we obtain the equation

$$E(N_4) = \tfrac{1}{2}[1 + E(N_2)] + \tfrac{1}{2} \cdot 1.$$

We know that $E(N_2) = 6$ and so $E(N_4) = 4$. This is the same answer as in Problem 2a. Hence the mean waiting time until repetition of the pattern (1 1) is the same as for (1 0).

1.5   Classical random walk I

In this book we shall discuss many problems concerning random walks; see Chapter 10 in particular. Here is a classical situation:

A particle performs a random walk on the integers of the x-axis, starting from the origin. At the time-points $t = 1, 2, \ldots$ the particle moves one step to the right or one step to the left with the same probability $\frac{1}{2}$. Such a walk is called *symmetric*. The walk stops when the particle arrives at one of the points $x = -a$ or $x = b$. Here a and b are positive integers. We then say that the particle is *absorbed* at one of these points (see Figure 1). We shall consider three problems for such a walk.

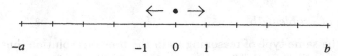

Fig. 1. Random walk with two absorbing barriers.

(a) *Probability of absorption*

We want to find the probability that the particle is absorbed at the point $x = -a$.

The answer is easier to find if we attack a more general problem. Suppose that, at a given moment, the particle is at the point $x = n$. Let p_n be the probability that, given this event, the particle is absorbed at the point $x = -a$.

It is realized that

$$p_n = \tfrac{1}{2}p_{n+1} + \tfrac{1}{2}p_{n-1}.$$

(Let the particle take one step, to the right or to the left.) Since the points $(n - 1, p_{n-1})$, (n, p_n), $(n + 1, p_{n+1})$ lie on a straight line, we conclude that

$$p_n = c_1 n + c_2.$$

The constants c_1, c_2 are determined by observing that $p_{-a} = 1$ and $p_b = 0$. A simple calculation then shows that

$$p_n = \frac{b - n}{a + b}.$$

By taking $n = 0$ we find that the probability is $b/(a + b)$ that the particle is absorbed in the point $x = -a$.

In a similar way, it is shown that the probability is $a/(a + b)$ that the particle is absorbed in the point $x = b$. (Incidentally, this proves that the walk cannot go on indefinitely, since the two probabilities have sum 1.)

(b) Expected number of steps until absorption

We want to determine the mean number of steps until the particle is absorbed. Again we solve a more general problem. Given that the particle is at the point $x = n$, let N_n be the time (=number of steps) until the particle is absorbed. The following difference equation holds:

$$E(N_n) = 1 + \tfrac{1}{2}E(N_{n+1}) + \tfrac{1}{2}E(N_{n-1}).$$

The general solution is

$$E(N_n) = -n^2 + c_1 n + c_2.$$

For $n = -a$ and $n = b$ we have $N_n = 0$. Using these boundary conditions we obtain

$$E(N_n) = (n + a)(b - n).$$

Taking $n = 0$ we find the final answer

$$E(N_0) = ab,$$

or in words: the expected time until absorption is ab.

(c) The ruin problem

We shall apply the results obtained above to the famous *ruin problem*.

Players A and B have a and b dollars, respectively. They repeatedly toss a fair coin. If heads appears, A gets \$1 from B; if tails appears, B gets \$1 from A. The game goes on until one of the players is ruined.

We want to determine the probability that A becomes ruined and find the mean number of tosses until the game ends.

The solutions are obtained from (a) and (b). The results of the successive tosses are equivalent to the behaviour of a particle performing a random walk along the x-axis starting from the origin. Let the particle take one step to the right if A wins, and one step to the left if B wins.

When the particle reaches the point $x = -a$, player A has lost his initial capital and is ruined. From (a) it follows that A becomes ruined with probability $b/(a + b)$. Likewise, B becomes ruined with probability $a/(a + b)$. Furthermore, it follows from (b) that the expected number of tosses until one of the players is ruined is equal to ab.

For example, if $a = 1$ and $b = 100$, player A is ruined with probability $100/101$ and B with probability $1/101$. The expected number of tosses of the game is 100, which is an astonishingly large number, in view of the fact that the probability is $\frac{1}{2}$ that the game ends after one toss!

This classical random walk is also discussed in Sections 10.2 and 16.2.

1.6   Number of walks until no shoes

A has a house with one front door and one back door. He places two pairs of walking shoes at each door. For each walk, he selects one door at random, puts on a pair of shoes, returns after a walk to a randomly chosen door, and takes off the shoes at the door. We want to determine the expected number of finished walks until A discovers that no shoes are available at the door he has selected for his next walk. The problem occurs in *The American Mathematical Monthly*, Problem E3043 (1984, p. 310 and 1987, p. 79).

We assume that at the start there are n pairs at each door. Consider a moment when there are i pairs of shoes available outside the front door, where $i = 0, 1, \ldots, 2n$. Let T_i be the number of completed walks until no pair is availabe when wanted.

First, let $i = 1, 2, \ldots, 2n - 1$. Considering what happens when one more walk is completed, we obtain the relation

$$E(T_i) = \tfrac{1}{4}[1 + E(T_{i-1})] + \tfrac{1}{2}[1 + E(T_i)] + \tfrac{1}{4}[1 + E(T_{i+1})]. \qquad (1)$$

Second, let $i = 0$. We then find

$$E(T_0) = \tfrac{1}{2} \cdot 0 + \tfrac{1}{4}[1 + E(T_1)] + \tfrac{1}{4}[1 + E(T_0)]$$

or

$$E(T_0) = \tfrac{1}{3}[2 + E(T_1)]. \qquad (2)$$

Third, for reasons of symmetry, we have the relation

$$E(T_i) = E(T_{2n-i}). \qquad (3)$$

Relation (1) is an inhomogeneous difference equation with the general solution

$$E(T_i) = c_1 + c_2 i - 2i^2,$$

where c_1 and c_2 are constants. Using relations (2) and (3), we obtain $c_1 = 2n, c_2 = 4n$. Hence,

$$E(T_i) = 2n + 4ni - 2i^2.$$

We are interested in the special case $n = 2, i = 2$. The mean is then 12. Hence, with two pairs of shoes at each door to begin with, A can perform 12 walks, on the average, before no shoes are available.

1.7   Banach's match box problem

Here we discuss a special version of the famous *Banach's match box problem. Stefan Banach* (1892–1945) was a Polish mathematician.

A person has, in each of his two pockets, a box with n matches. Now and then he takes a match from a randomly chosen box until he finds the selected box empty. Find the expectation of the number, R, of remaining matches in the other box.

(a) *'Translation' to a coin problem*

Choosing a match from box 1 or box 2 is equivalent to flipping a coin. The tosses go on until heads has come up $n + 1$ times or tails has come up $n + 1$ times. (The chosen box is found empty after n matches have been taken from it and a further attempt has been made.) Let N be the number of tosses required. This number corresponds to the sum of (i) the n matches in the box plus 1 for the extra attempt and (ii) the $n - R$ matches taken from the other box. Thus we have the relation

$$N = 2n + 1 - R,$$

or, rearranging and taking the average,

$$E(R) = 2n + 1 - E(N). \tag{1}$$

Thus we need $E(N)$.

(b) *Solution to the coin problem*

The results of the successive tosses can be visualized as a random walk in the first quadrant of an (x, y)-system of points (see Figure 1). The walk

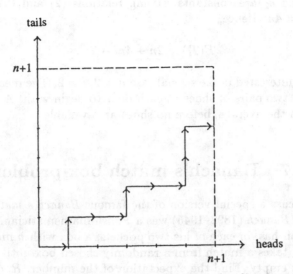

Fig. 1. Random walk with stopping lines.

stops when either the line $x = n + 1$ or the line $y = n + 1$ is attained. The number of steps is the rv N mentioned in (a).

We now determine the probability function of N, $p_k = P(N = k)$, for $k = n + 1, n + 2, \ldots, 2n + 1$. It is seen that $N = k$ if either

1. n of the first $k - 1$ tosses come up heads, as well as the kth toss or

2. n of the first $k - 1$ tosses come up tails, as well as the kth toss.

By symmetry, these two events occur with the same probability. Hence by well-known properties of the binomial distribution

$$p_k = 2\binom{k-1}{n}\left(\frac{1}{2}\right)^{k-1}\frac{1}{2} = \binom{k-1}{n}\left(\frac{1}{2}\right)^{k-1}$$

for $k = n + 1, n + 2, \ldots, 2n + 1$.

In order to find the expected value of N we use the recursive relation

$$\frac{p_k}{p_{k+1}} = \frac{2(k-n)}{k} \tag{2}$$

for $k = n + 1, n + 2, \ldots, 2n$, noting that

$$p_{n+1} = \left(\frac{1}{2}\right)^n; \qquad p_{2n+1} = \binom{2n}{n}\left(\frac{1}{2}\right)^{2n}.$$

Formula (2) can be rewritten as

$$2(k+1)p_{k+1} = kp_k + 2(n+1)p_{k+1}.$$

Summing for $k = n + 1, n + 2, \ldots, 2n$, we find

$$2E(N) - 2(n + 1)p_{n+1} = E(N) - (2n + 1)p_{2n+1} + 2(n + 1)(1 - p_{n+1}).$$

After reduction we obtain

$$E(N) = 2(n + 1) - (2n + 1)\binom{2n}{n}\left(\frac{1}{2}\right)^{2n}. \tag{3}$$

(c) *Solution to the match box problem*

Inserting (3) into (1) we obtain

$$E(R) = (2n + 1)\binom{2n}{n}\left(\frac{1}{2}\right)^{2n} - 1,$$

which gives the exact answer to the original problem.

When n is large we use Stirling's formula (see 'Symbols and formulas' at the beginning of the book) and after some rather tedious calculation, we obtain

$$E(R) \approx 2\sqrt{\frac{n}{\pi}} - 1 + \frac{3}{4\sqrt{n\pi}}.$$

For example, if there are $n = 50$ matches in each box from the beginning, there are, on average, about 7 matches left in one box when the other is found empty. (A more accurate value is 7.039; both the exact formula and the approximation give this value.)

We return to the coin problem at the end of Section 10.6.

1.8   The generous king

The king of a country with m inhabitants is generous enough to give each citizen a gold coin as a Christmas gift. One year, gambling becomes popular among his people, and, therefore, he wishes to replace the usual procedure by doing the following:

All m citizens line up in random order in front of the Royal Palace. The king comes out on to a balcony and tosses k gold coins in the air. When a coin reaches the ground it comes up heads or tails with the same probability $\frac{1}{2}$. The first person in the queue receives as a gift all coins showing tails and returns the others to the king; he then leaves the queue. The king tosses the remaining coins; the second person obtains those which come up tails and returns the rest, and so on. The procedure stops at the mth toss, or earlier if there are no coins left to distribute. Note that, since the citizens wait in random order, the procedure is fair, in spite of the fact that the first citizen receives, on the average, $k/2$ coins, the second $k/4$, and so on.

Example

Suppose that $m = 5$ and $k = 6$. We may then obtain, for example:

Toss	Heads	Tails	Sum
1	3	3	6
2	2	1	3
3	1	1	2
4	0	1	1

The five citizens in the queue receive 3, 1, 1, 1 and 0 coins, respectively.

We shall now help the king to choose a suitable number of coins for his gift. In an interview, the king expresses the following wish: The probability that the last person in the queue wins a coin must be at least $\frac{1}{2}$.

Let N_k be the number of people given the opportunity to win a coin. In other words, the king performs exactly N_k tosses before there are no coins left to distribute. Clearly, N_k is an rv which attains the values $1, 2, \ldots, m$. We want to find k such that

$$P(N_k \geq m) \geq \tfrac{1}{2}. \tag{1}$$

Let us determine the probability

$$p_n = P(N_k > n).$$

For that purpose, number the coins from 1 to k. Let B_i denote the event that coin i comes up heads in each of the first n tosses. The tosses are independent, hence

$$P(B_i) = \left(\tfrac{1}{2}\right)^n. \tag{2}$$

It is seen that $N_k > n$ if at least one of the k coins comes up heads in each of the first n tosses, for only then does the king have something to distribute in these tosses. Hence $N_k > n$ if at least one of the B_i's occur, and so by (2) (since the B's are independent)

$$p_n = P(N_k > n) = 1 - \left[1 - \left(\tfrac{1}{2}\right)^n\right]^k. \tag{3}$$

We now turn our attention to (1), where we have to find k. From (3) we obtain the relation

$$1 - \left[1 - \left(\tfrac{1}{2}\right)^{m-1}\right]^k \geq \tfrac{1}{2}.$$

We want to find the smallest k such that this inequality is satisfied. After some manipulations, we obtain

$$k \geq \frac{-\ln 2}{\ln[1 - 1/2^{m-1}]} \approx \ln 2 \cdot 2^{m-1},$$

where the approximation holds for large m.

The result shows, for example, that if $m = 1000$ (a moderate population indeed), the king should start with about 10^{300} coins! After hearing this, he scraps the whole idea.

2

Basic probability theory I

In this chapter, we visit the vaguely defined area of basic probability theory; in our conception of the world of probability this area includes elementary theory for rv's. Ideas are put forward concerning conditional probability, exchangeability, combination of events and zero–one rv's. Do not forget the last section about the 'zero–one idea', which provides a powerful tool in many otherwise awkward situations. A good general reference for basic discrete probability is Feller (1968); another is Moran (1968).

2.1 Remarkable conditional probabilities

Let A and B be two events. We seek the conditional probability that both A and B occur, under two different conditions.

Condition 1. It is known that A has occurred

Denoting the conditional probability that AB occurs by P_1, we obtain

$$P_1 = P(AB|A) = \frac{P(AB \cup A)}{P(A)} = \frac{P(AB)}{P(A)}. \tag{1}$$

Condition 2. It is known that $A \cup B$ has occurred

As the reader will know, this means that at least one of the events A and B has occurred. We seek the conditional probability P_2 of AB given this event. We have

$$P_2 = P(AB|A \cup B) = \frac{P[AB \cap (A \cup B)]}{P(A \cup B)}.$$

Now $AB \subseteq A \cup B$, and so the numerator reduces to $P(AB)$. Hence, we have

$$P_2 = \frac{P(AB)}{P(A \cup B)}. \tag{2}$$

Let us now suppose that B is not totally contained in A. (Otherwise $A \cup B = A$, which makes our question uninteresting.) We then have

$$P(A^*B) > 0, \tag{3}$$

where A^* is the complement of A. Under this condition we obtain

$$P(A \cup B) = P(A) + P(A^*B) > P(A).$$

Therefore, as a consequence of (1) and (2) we infer that

$$P_2 < P_1. \tag{4}$$

This inequality seems remarkable, at least to the untrained eye. The knowledge that $A \cup B$ has taken place, that is, that A or B or both have occurred, makes the conditional probability that AB happens smaller than when we know that A has happened.

Problem 1

In a lottery there are 100 tickets and among them 2 prize tickets, A and B. Adam buys four tickets.

   He first tells Friend 1 of his purchase and that he obtained the prize A. The friend asks himself: What is the probability that Adam won both prizes? We have

$$P(A) = \binom{1}{1}\binom{99}{3} \Big/ \binom{100}{4} = \frac{4}{100},$$

$$P(AB) = \binom{2}{2}\binom{98}{2} \Big/ \binom{100}{4} = \frac{4 \cdot 3}{100 \cdot 99},$$

which in view of (1) leads to

$$P_1 = \frac{1}{33}.$$

   Adam then tells Friend 2 that the purchase gave him at least one of the prizes. Friend 2 puts the same question as Friend 1. We find

$$P(A \cup B) = P(A) + P(B) - P(AB)$$
$$= \frac{4}{100} + \frac{4}{100} - \frac{4 \cdot 3}{100 \cdot 99} = \frac{13}{165}.$$

It follows from (2) that

$$P_2 = \frac{1}{65}.$$

Thus P_2 is about the half of P_1, which seems remarkable.

Problem 2

A bridge player announces that, among the 13 cards which he has received,

   a. one is the ace of hearts

   b. there is at least one ace.

The conditional probability that he has received more than one ace is $11,686/20,825 \approx 0.5611$ in a and $5,359/14,498 \approx 0.3696$ in b. We invite the reader to show this.

2.2   Exchangeability I

Exchangeability is an important concept in probability theory. A succinct definition will be given and illustrated by three examples.

Consider a finite sequence of events $A_1, A_2, \ldots, A_n$ or an infinite sequence $A_1, A_2, \ldots$. Assume that, for any given integer k, the probability

$$P(A_{j_1} A_{j_2} \cdots A_{j_k})$$

is the same for any k indices $j_1 < j_2 < \cdots < j_k$. (Of course, when the sequence is finite, $k \leq n$.) The events are then said to be *exchangeable*. Another good term is symmetrically dependent, but it is seldom used nowadays.

Example 1. *Four exchangeable events*

Consider four events A_1, A_2, A_3, A_4. Write for i, j, k equal to $1, 2, 3, 4$

$$p_i = P(A_i),$$
$$p_{ij} = P(A_i A_j), \ i < j,$$
$$p_{ijk} = P(A_i A_j A_k), \ i < j < k.$$

In the general case, the four p_i's may all be different, as well as the six p_{ij}'s and the four p_{ijk}'s. Now suppose that all p_i's are equal, and, say, equal to p_1, that all p_{ij}'s are equal to p_{12} and all p_{ijk}'s are equal to p_{123}. The events are then exchangeable. This property makes life simpler for the probabilist; in this example he needs only four quantities for defining all fifteen probabilities involved.

Example 2. *Random permutation*

Consider a random permutation of the integers $1, 2, \ldots, n$. Let A_i be the event that the number i occupies place i in the random permutation, where $i = 1, 2, \ldots, n$. The A's constitute a finite sequence of exchangeable events. (This is obvious for reasons of symmetry.)

Example 3. Drawings without replacement

An urn contains a white and b black balls. Balls are drawn, one at a time at random without replacement, until the urn is empty. Let A_i be the event that 'the ith ball drawn is white', $i = 1, 2, \ldots, a + b$. The A_i's constitute a finite sequence of exchangeable events.

References: Feller (1971, p. 228), Johnson and Kotz (1977, p. 97).

Here is a problem for the reader. An urn contains a white balls and b black balls. Draw the balls one at a time at random without replacement. Show that the probability that all black balls are drawn before the last white ball is $a/(a + b)$.

2.3   Exchangeability II

The theory of exchangeability is largely due to the Italian probabilist and statistician *Bruno de Finetti* (1906–1985). An important theorem, *de Finetti's theorem*, will be presented in this section. It holds for *infinite* sequences of exchangeable events.

In order to describe the theorem we make the following construction: Consider a sequence of n independent trials, where, at a given trial, an event H occurs with probability p. Let A_i be the event that H occurs at the ith trial. Then $A_1, \ldots, A_n$ are independent events. Let X be the number of events that occur among $A_1, \ldots, A_n$. The rv X has a binomial distribution:

$$P(X = k) = \binom{n}{k} p^k (1 - p)^{n-k}, \tag{1}$$

where $k = 0, 1, \ldots, n$.

We shall now extend the description from independent to exchangeable events. Suppose that p is no longer constant but an rv taking values over the interval $(0, 1)$. Its distribution can be continuous or discrete; for brevity we assume that it is continuous, but the discrete case is entirely analogous. What happens now to the probabilities involved?

Let $f(p)$ be the density function of p. We find

$$P(A_i) = \int_0^1 p f(p) \, dp, \tag{2}$$

$$P(A_i A_j) = \int_0^1 p^2 f(p) \, dp \qquad (i \neq j) \tag{3}$$

and so on. Evidently, the A_i's constitute a finite sequence of exchangeable events, for any choice of density function.

The density function $f(p)$ is often called the *prior distribution* or, more briefly, the *prior*.

We obtain from (1) the important expression

$$P(X = k) = \int_0^1 \binom{n}{k} p^k (1-p)^{n-k} f(p)\, dp, \qquad (4)$$

where $k = 0, 1, \ldots, n$. This is the probability function of X.

After these preparations we are ready for de Finetti's theorem. Consider any infinite sequence $A_1, A_2, \ldots$ of exchangeable events. According to the theorem, the finite subsequences $A_1, A_2, \ldots, A_n$ of such a sequence can always be obtained in the way described by (4): *Start with independent events and integrate over the prior.* (Any infinite sequence has its own prior, which has to be determined in each special case.)

On the other hand, if we assign a fixed value to p, the A_i's become independent and (1) holds. Expressed otherwise, exchangeable A's taken from an infinite sequence are conditionally independent, given p.

De Finetti's theorem was first published in 1930 in an Italian journal. For proofs of the theorem, see Johnson and Kotz (1977, p. 103) or Heath and Sudderth (1976). An application is given in Section 11.3.

2.4   Combinations of events I

Let $A_1, A_2, \ldots, A_n$ be events defined on the same sample space. In the history of probability, many authors have posed problems leading to the determination of the probabilities

$$p_n(k) = P(\text{exactly } k \text{ of the events occur})$$

and

$$P_n(k) = P(\text{at least } k \text{ of the events occur}),$$

where $1 \leq k \leq n$.

Such probabilities were studied by *Abraham de Moivre* (1667–1754) in the first edition of his famous book *The Doctrine of Chances* from 1718; the second and third editions were published in 1738 and 1756. He understood their general importance, but formulated his results with specific examples.

(a) *Exchangeable events*

In de Moivre's problem, the events are exchangeable; that is, the probabilities $P(A_{i_1} A_{i_2} \cdots A_{i_r})$ are the same for all r-tuples $(i_1, i_2, \ldots, i_r)$ and any r; see Section 2.2. For example, if $n = 3$ and $r = 2$, we shall have $P(A_1 A_2) = P(A_1 A_3) = P(A_2 A_3)$.

By a clever application of formulas like

$$P(A_1^* A_2) = P(A_2) - P(A_1 A_2),$$
$$P(A_1^* A_2 A_3) = P(A_2 A_3) - P(A_1 A_2 A_3)$$

de Moivre found, for example, that

$$P(A_1 A_2 A_3^* A_4^*) = P(A_1 A_2) - 2P(A_1 A_2 A_3) + P(A_1 A_2 A_3 A_4).$$

Multiplying this expression by $\binom{4}{2}$, the number of ways to choose two A's out of four, we obtain $p_4(2)$. (As usual, an asterisk denotes complement.) The general formulas for $p_n(k)$ and $P_n(k)$ were stated, but not proved, by de Moivre:

$$p_n(k) = \binom{n}{k} \sum_{i=k}^{n} (-1)^{i-k} \binom{n-k}{i-k} P(A_1 \cdots A_i), \tag{1}$$

$$P_n(k) = \sum_{i=k}^{n} (-1)^{i-k} \binom{i-1}{i-k} \binom{n}{i} P(A_1 \cdots A_i). \tag{2}$$

(b) *The general case*

For general events, probability had been around a long time before formulas for $p_n(k)$ and $P_n(k)$ became known. These formulas are the following:

$$p_n(k) = \sum_{i=k}^{n} (-1)^{i-k} \binom{i}{k} S_i, \tag{3}$$

$$P_n(k) = \sum_{i=k}^{n} (-1)^{i-k} \binom{i-1}{k-1} S_i. \tag{4}$$

Here

$$S_1 = \sum_i P(A_i); \quad S_2 = \sum_{i<j} P(A_i A_j); \quad S_3 = \sum_{i<j<k} P(A_i A_j A_k) \tag{5}$$

and so on.

The special case $k = 1$ of (4) is important; it is a slightly disguised version of the well-known addition formula

$$P(A_1 \cup A_2 \cup \cdots \cup A_n) = S_1 - S_2 + S_3 - \cdots + (-1)^{n-1} S_n. \tag{6}$$

This formula is often called the *inclusion–exclusion formula*. Proofs of (3) and (4) are given in Section 3.6.

Finally, we mention the following useful inequalities, the first of which is often called *Boole's inequality*:

$$P(A_1 \cup \cdots \cup A_n) \le S_1,$$
$$P(A_1 \cup \cdots \cup A_n) \ge S_1 - S_2,$$
$$P(A_1 \cup \cdots \cup A_n) \le S_1 - S_2 + S_3$$

and so on. A closely related inequality is

$$P(A_1 A_2 \cdots A_n) \ge 1 - \sum_{i=1}^{n} P(A_i^*).$$

Applications of some of the formulas given in this section are found in Section 2.5.

We finish the section with a problem involving exchangeable events. There are four parking places at each of four streets surrounding a block. Eight cars park at random in the sixteen available places. Show, for example by using (6), that the probability that at least one car has parked at each street is equal to $1{,}816/2{,}145 \approx 0.8466$.

2.5   Problems concerning random numbers

Choose a 6-digit decimal random number, that is, one of the numbers $000000, 000001, \ldots, 999999$ chosen at random.

(a) *Problem 1*

Find the probability that at least one of the digits $0, 1, \ldots, 9$ appears exactly twice.

Let P be the probability of the event we are interested in and A_i the event that the integer i appears exactly twice, where $i = 0, 1, \ldots, 9$. We obtain by the binomial and the multinomial distributions

$$P_1 = P(A_0) = \frac{6!}{2!4!} \, 0.1^2 0.9^4 = \frac{98{,}415}{10^6},$$

$$P_2 = P(A_0 A_1) = \frac{6!}{(2!)^3} \, 0.1^2 0.1^2 0.8^2 = \frac{5{,}760}{10^6},$$

$$P_3 = P(A_0 A_1 A_2) = \frac{6!}{(2!)^3 (0!)} \, 0.1^2 0.1^2 0.1^2 0.7^0 = \frac{90}{10^6}.$$

Inserting these expressions in (6) of the preceding section, we obtain

$$P = \binom{10}{1} P_1 - \binom{10}{2} P_2 + \binom{10}{3} P_3 = \frac{2{,}943}{4{,}000} \approx 0.7358.$$

Alternatively, we could have used formula (2).

(b) *Problem 2*

Find the probability that at least two of the digits $0, 1, \ldots, 9$ appear exactly once.

Let P be the probability of the event studied and A_i the event that the integer i appears exactly once. We have by the multinomial distribution

$$P_2 = P(A_0 A_1) = \frac{6!}{(1!)^2 4!} 0.1^2 0.8^4,$$

$$P_3 = P(A_0 A_1 A_2) = \frac{6!}{(1!)^3 3!} 0.1^3 0.7^3,$$

$$P_4 = P(A_0 \cdots A_3) = \frac{6!}{(1!)^4 2!} 0.1^4 0.6^2,$$

$$P_5 = P(A_0 \cdots A_4) = \frac{6!}{(1!)^5 1!} 0.1^5 0.5^1,$$

$$P_6 = P(A_0 \cdots A_5) = \frac{6!}{(1!)^6 0!} 0.1^6 0.4^0.$$

Inserting these expressions in formula (2) of the preceding section, and letting $k = 2$, we obtain (note that $P_7 = \cdots = P_{10} = 0$)

$$P = \sum_{i=2}^{10} (-1)^{i-2} \binom{i-1}{i-2} \binom{10}{i} P_i = \frac{1,179}{1,250} \approx 0.9432.$$

2.6   Zero–one random variables I

We shall describe a marvellous device, which is of great help to the probabilist: *zero–one* rv's. Other names are *indicator* rv's and *Bernoulli* rv's. As the name suggests, a zero–one rv only takes the values 0 and 1. For example, a 1 may indicate that a certain event happens and a 0 that it does not.

More generally, if $A_1, \ldots, A_n$ are events, we may introduce zero–one rv's $U_1, \ldots, U_n$ such that $U_i = 1$ if A_i happens and $U_i = 0$ if A_i does not. Then the sum

$$X_n = U_1 + U_2 + \cdots + U_n \tag{1}$$

is the total number of events that happen.

We can see that

$$E(U_i) = 1 \cdot P(A_i) + 0 \cdot [1 - P(A_i)] = P(A_i). \tag{2}$$

Inserting this in (1) we obtain

$$E(X_n) = \sum_{i=1}^{n} E(U_i) = \sum_{i=1}^{n} P(A_i). \tag{3}$$

Therefore, if the individual probabilities are known, we can easily find the mean of the total count. Note that this works even if the events A_i are dependent, which is wonderful.

As $U_i^2 \equiv U_i$, the variance of U_i is given by

$$Var(U_i) = E(U_i^2) - [E(U_i)]^2 = P(A_i)[1 - P(A_i)],$$

and for $i \neq j$ the covariance of U_i and U_j is given by

$$Cov(U_i, U_j) = E(U_i U_j) - E(U_i)E(U_j) = P(A_i A_j) - P(A_i)P(A_j).$$

Hence we get the following expression for the variance of X_n:

$$Var(U_1 + U_2 + \cdots + U_n) = \sum_{i=1}^{n} Var(U_i) + 2 \sum_{1 \leq i < j \leq n} Cov(U_i, U_j)$$

$$= \sum_{i=1}^{n} P(A_i)[1 - P(A_i)] + 2 \sum_{1 \leq i < j \leq n} [P(A_i A_j) - P(A_i)P(A_j)]. \tag{4}$$

In particular, if the A_i's are independent events, (4) simplifies to

$$Var(X_n) = \sum_{i=1}^{n} P(A_i)[1 - P(A_i)]. \tag{5}$$

We shall mention two special cases when the U_i's are independent.

(a) *Binomial distribution*

If each U_i takes values 1 and 0 with probabilities p and $1 - p$, respectively, we realize that each U_i is $Bin(1, p)$. As a consequence, X_n has a binomial distribution $Bin(n, p)$ with probability function

$$P(X_n = k) = \binom{n}{k} p^k (1 - p)^{n-k},$$

where $k = 0, 1, \ldots, n$. We then have by (3) and (5)

$$E(X_n) = np; \quad Var(X_n) = np(1 - p).$$

(b) *Poisson–binomial distribution*

Suppose that the U_i's have unequal p's so that U_i is 1 or 0 with probabilities p_i and $1 - p_i$, respectively. Then the mean and variance of X_n in (1) are given by

$$E(X_n) = \sum_{i=1}^{n} p_i; \quad Var(X_n) = \sum_{i=1}^{n} p_i(1 - p_i).$$

The rv X_n is said to have a *Poisson–binomial* distribution. The probabilities $P(X_n = k)$ will be derived in Section 3.5; see (4) in that section.

Example 1

One hundred students were attending a lecture in probability theory sitting on chairs numbered 1 to 100. The lecturer said: Imagine that I rank you according to your age so that the youngest gets number 1, the next youngest number 2, etc. How many will have the same number for the rank and the chair?

In order to answer the question we have to make assumptions on how the students are seated. Assume that the seating is independent of the age of the students, which seems reasonable. We also assume that there is a strict ranking from 1 to 100 without ties. Let U_i be 1 if the ith student sits on chair i and 0 otherwise. Then the U_i's are zero–one rv's with common $p = 1/100$. Their sum X_{100} is the number of students with the same rank and chair numbers. We have

$$E(X_{100}) = 100 \cdot \frac{1}{100} = 1.$$

It might come as a surprise that the expected number is *one* for any number of students. Here, the U_i's are not independent, so X_{100} has *not* a binomial distribution.

Example 2

A fair coin is tossed twelve times. Denoting heads by H and tails by T, the result can be written as a sequence of twelve H's and T's. Let X be the number of HHH that can be read in this sequence. (Example: When the sequence is $TTHTHHHHHTHHH$, we have $X = 3$.) Using zero–one rv's it is easy to show that X has expectation 5/4.

3

Basic probability theory II

Our presentation of problems and snapshots from basic probability theory is continued. A trick is shown for determining expectations, and we then discuss, among other things, probability generating functions and factorial generating functions.

For further reading, see Feller (1968) and Moran (1968).

3.1 A trick for determining expectations

Tricks are sometimes used when calculating probabilities; they sweeten the life of the probabilist.

Let X be an rv assuming the values $0, 1, 2, \ldots$, or a finite number of these values. The *expectation* of X is defined by

$$E(X) = \sum_{k=0}^{\infty} k P(X = k).$$ (1)

Other names are *expected value* and *mean*.

Now comes the trick. According to a well-known theorem on positive series, the sum of such a series is unchanged if the order of summation is changed. Therefore we can write

$$
\begin{aligned}
E(X) &= P(X = 1) + [P(X = 2) + P(X = 2)] \\
&\quad + [P(X = 3) + P(X = 3) + P(X = 3)] + \cdots \\
&= [P(X = 1) + P(X = 2) + \cdots] + [P(X = 2) + P(X = 3) + \cdots] \\
&\quad + [P(X = 3) + P(X = 4) + \cdots] + \cdots .
\end{aligned}
$$

More succinctly, we have proved that

$$E(X) = \sum_{k=0}^{\infty} P(X > k).$$ (2)

Sometimes this expression is simpler to evaluate than (1). Also (2) is convenient to use when the probabilities $P(X > k)$, $k = 0, 1, \ldots$, are simpler to calculate than the probabilities $P(X = k)$. Formula (2) will be frequently used in the sequel.

Also the second moment can be expressed in terms of the probabilities $P(X > k)$:

$$E(X^2) = \sum_{k=0}^{\infty}(2k+1)P(X > k).$$

We leave the proof to the reader.

3.2 Probability generating functions

We often use probability generating functions in this book, and it is essential that the reader is familiar with this useful tool.

Let X be an rv taking the values $0, 1, 2, \ldots$. The *probability generating function* $G_X(s)$ of X is defined by

$$G_X(s) = E(s^X). \tag{1}$$

Here s is a variable in the ordinary mathematical sense. This function can be used for different purposes, three of which will be mentioned here:

(a) *Determination of probabilities*

We have from the definition (1) that

$$G_X(s) = \sum_{k=0}^{\infty} s^k P(X = k). \tag{2}$$

We note in passing that $G_X(1) = 1$. If $G_X(s)$ is known, we may expand this function in a power series and, in this way, by identification obtain the probability function $P(X = k)$ for $k = 0, 1, \ldots$. This is a common procedure in probability theory.

(b) *Determination of moments*

Taking in (2) the first derivative with respect to s and setting $s = 1$, we find

$$G'_X(1) = \sum_{k=0}^{\infty} kP(X = k) = E(X). \tag{3}$$

More generally, by taking the jth derivative we obtain the jth *descending factorial moment* of X:

$$G_X^{(j)}(1) = E[X(X-1)\cdots(X-j+1)]. \tag{4}$$

From the factorial moments, the ordinary central moments $E[(X-\mu)^j]$ can be obtained by forming suitable linear combinations; here $\mu = E(X)$.

(c) *Determination of probabilities of sums of rv's*

Let $Z = X + Y$, where X and Y are independent rv's. The independence gives

$$G_Z(s) = E(s^{X+Y}) = E(s^X)E(s^Y) = G_X(s)G_Y(s). \tag{5}$$

This is an elegant result: A summation of independent rv's corresponds to a multiplication of their probability generating functions. By expanding the product in a power series we may determine the probability function $P(Z = k)$, $k = 0, 1, \ldots$, of the sum Z.

Example. Binomial distribution

If X has a binomial distribution $\mathrm{Bin}(n, p)$, we may write X as a sum,

$$X = U_1 + U_2 + \cdots + U_n,$$

of zero–one rv's; compare Section 2.6. Here each U_i is 1 or 0 with probability p and $q = 1 - p$, respectively. We have for any U_i

$$G_{U_i}(s) = E(s^{U_i}) = s \cdot P(U_i = 1) + 1 \cdot P(U_i = 0) = ps + q.$$

By an obvious generalization of (5) to n independent rv's we obtain the probability generating function

$$G_X(s) = (ps + q)^n$$

of the rv X. Taking the first derivative we obtain

$$G'_X(s) = np\,(ps + q)^{n-1}.$$

Putting $s = 1$ we have the familiar result

$$E(X) = G'_X(1) = np.$$

By taking further derivatives, we find that the jth descending factorial moment of X is given by

$$E[X(X-1)\cdots(X-j+1)] = G_X^{(j)}(1) = n(n-1)\cdots(n-j+1)p^j\,.$$

Here is a problem for the reader. Five urns contain the following numbers of white and black balls (the numbers of white balls are given first): $(1, 2)$, $(2, 2)$, $(3, 1)$, $(2, 5)$, $(3, 2)$. One ball is drawn at random from each urn. Derive the probability generating function of the total number X of white balls drawn and use this function for showing that $P(X = 2) = 283/840 \approx 0.3369$.

An excellent book on generating functions is by Wilf (1990).

3.3   People at the corners of a triangle

At the time-point $t = 0$ one at each corner. At the moments $t = 1, 2, \ldots$ each person moves to one of the other two corners chosen at random. Let M be the waiting time (the number of steps) until the three people meet at the same corner. Find the probability generating function of the rv M and use this function for determining the expectation and the variance of M.

We may solve this problem as follows: Besides M, we need the waiting time N until the three people meet at the same corner given that exactly two of them are at the same corner.

Let us now start at $t = 0$. At $t = 1$, eight different cases can occur with equal probability $1/8$, depending on how the people move. In two cases the three people are again at different corners, and in six cases exactly two of them are at the same corner. Hence

$$M = \begin{cases} 1 + M' & \text{with probability } 1/4 \\ 1 + N & \text{with probability } 3/4, \end{cases}$$

where M' has the same distribution as M.

Now suppose that exactly two people are at the same corner. When they move, there are again eight different cases. In one of these cases all people meet at the same corner; in five cases there are again two people at the same corner and in two cases there is one person at each corner. Hence

$$N = \begin{cases} 1 & \text{with probability } 1/8 \\ 1 + N' & \text{with probability } 5/8 \\ 1 + M & \text{with probability } 1/4, \end{cases}$$

where N' has the same distribution as N.

Denote by

$$P(s) = E(s^M); \quad Q(s) = E(s^N)$$

the probability generating functions of M and N, respectively. From the relations derived above it follows that

$$P(s) = \frac{s}{4}P(s) + \frac{3s}{4}Q(s),$$

$$Q(s) = \frac{s}{8} + \frac{5s}{8}Q(s) + \frac{s}{4}P(s).$$

Solving this system of equations in $P(s)$ and $Q(s)$ with respect to $P(s)$, we find

$$P(s) = \frac{3s^2}{32 - 28s - s^2}.$$

Differentiating this probability generating function with respect to s and setting $s = 1$, after some calculation we obtain that $P'(1) = 12$ and $P''(1) = 728/3$. Hence, by the properties of probability generating functions

$$E(M) = P'(1) = 12$$

and

$$Var(M) = P''(1) + P'(1) - [P'(1)]^2 = \frac{332}{3}.$$

The probability function $p_m = P(M = m)$ can be obtained from the expression for $P(s)$. We have

$$p_m = \frac{3}{4\sqrt{57}} \left[\left(\frac{1}{s_1} \right)^{m-1} - \left(\frac{1}{s_2} \right)^{m-1} \right],$$

where $m \geq 2$, $s_1 = -14 + 2\sqrt{57}$ and $s_2 = -14 - 2\sqrt{57}$. The derivation is left to the reader.

3.4   Factorial generating functions

The factorial generating function is closely related to the probability generating function.

Let X assume values $0, 1, 2, \ldots$ and have the probability generating function $G_X(s) = E(s^X)$. We saw in Section 3.2 that the jth descending factorial moment can be obtained from this function. In fact, we proved that by differentiating the probability generating function j times and taking $s = 1$ we obtain

$$E[X(X - 1) \cdots (X - j + 1)] = G_X^{(j)}(1).$$

As a consequence, we may expand $G_X(s)$ in a power series about the point $s = 1$:

$$G_X(s) = \sum_{j=0}^{\infty} E\left[\binom{X}{j} \right] (s - 1)^j . \tag{1}$$

Here

$$E\left[\binom{X}{j} \right] = E[(X(X - 1) \cdots (X - j + 1)]/j!$$

is called the jth *binomial moment* of X. We now take $s = 1 + t$. The resulting function of t,

$$H_X(t) = G_X(1 + t) = E[(1 + t)^X],$$

is called the *factorial generating function* of the rv X. It follows from (1) that

$$H_X(t) = \sum_{j=0}^{\infty} E\left[\binom{X}{j}\right] t^j. \tag{2}$$

Hence we may obtain the binomial (or factorial) moments by developing the function $H_X(t)$ in a power series about the point $t = 0$; this explains the name of the function.

The probabilities $P(X = k)$ can be obtained 'backwards' in the following way. As seen from (1), we may write

$$G_X(s) = \sum_{j=0}^{\infty} E\left[\binom{X}{j}\right](s-1)^j$$

$$= \sum_{j=0}^{\infty} E\left[\binom{X}{j}\right] \sum_{k=0}^{j} \binom{j}{k}(-1)^{j-k} s^k$$

$$= \sum_{k=0}^{\infty} s^k \sum_{j=k}^{\infty}(-1)^{j-k}\binom{j}{k}E\left[\binom{X}{j}\right].$$

As

$$G_X(s) = \sum_{k=0}^{\infty} s^k P(X = k)$$

we obtain the inversion formula

$$P(X = k) = \sum_{j=k}^{\infty}(-1)^{j-k}\binom{j}{k}E\left[\binom{X}{j}\right]. \tag{3}$$

Here is a problem for the reader. If X has a geometric distribution with probability function $P(X = k) = pq^k$ for $k = 0, 1, \ldots$, then

$$E\left[\binom{X}{j}\right] = \left(\frac{q}{p}\right)^j.$$

Show this.

3.5   Zero–one random variables II

We shall return to the zero–one idea already praised in Section 2.6. Let X be a sum of n independent or dependent rv's

$$X = U_1 + U_2 + \cdots + U_n,$$

where U_i is 1 or 0 with probability p_i and $1 - p_i$, respectively. We shall solve four problems concerning the rv X.

(a) *Factorial generating function*

The factorial generating function $H_X(t) = E[(1+t)^X]$ may be obtained by a trick. It is seen that

$$(1+t)^{U_i} \equiv 1 + U_i t.$$

Hence $H_X(t)$ is given by

$$H_X(t) = E[(1+t)^{\Sigma U_i}] = E[(1+t)^{U_1} \cdots (1+t)^{U_n}]$$
$$= E[(1+U_1 t) \cdots (1+U_n t)].$$

Writing the right side as a polynomial in t, we find

$$H_X(t) = 1 + \left[\sum_i E(U_i)\right] t + \left[\sum_{i<j} E(U_i U_j)\right] t^2$$
$$+ \left[\sum_{i<j<k} E(U_i U_j U_k)\right] t^3 + \cdots + E(U_1 \cdots U_n) t^n.$$

In a condensed form this can be written

$$H_X(t) = \sum_{j=0}^{n} S_j\, t^j, \tag{1}$$

where $S_0 = 1$ and

$$S_j = \sum_{i_1 < \cdots < i_j} E(U_{i_1} \cdots U_{i_j}). \tag{2}$$

This expression does not look very promising. However, we can use it for three good purposes:

(b) *Binomial moments*

We want to find the jth *binomial moment*

$$E\left[\binom{X}{j}\right]$$

for $j = 1, 2, \ldots, n$. Comparing (1) above with (2) in the preceding section we conclude that

$$E\left[\binom{X}{j}\right] = S_j; \tag{3}$$

here S_j is defined in (2) above. For example, we have

$$E\left[\binom{X}{2}\right] = \sum_{i<j} E(U_i U_j).$$

(c) Probability function

We want to determine an explicit expression for the probability function $P(X = k)$ for $k = 0, 1, \ldots, n$. This seems difficult in view of the definition of X. However, the tools we now have at our disposal provide the solution. In view of (3) above we obtain from (3) in the preceding section the answer

$$P(X = k) = \sum_{j=k}^{n} (-1)^{j-k} \binom{j}{k} S_j. \tag{4}$$

This is an elegant result, but the S_j's are not always easy to determine.

(d) Distribution/survival function

Consider the distribution function of X, $F_X(k) = P(X \le k)$, or more conveniently the survival function $1 - F_X(k) = P(X > k)$. The generating function of $P(X > k)$ can be written successively (remembering that $X \le n$) as

$$\sum_{k=0}^{n-1} s^k P(X > k) = \sum_{k=0}^{n-1} \sum_{j=k+1}^{n} P(X = j)s^k = \sum_{j=1}^{n} P(X = j) \sum_{k=0}^{j-1} s^k$$

$$= \sum_{j=1}^{n} P(X = j) \frac{s^j - 1}{s - 1} = \frac{G_X(s) - 1}{s - 1}.$$

Since $G_X(s) = H_X(s - 1)$, we obtain by (1)

$$\frac{G_X(s) - 1}{s - 1} = \sum_{j=1}^{n} S_j (s - 1)^{j-1} = \sum_{j=0}^{n-1} S_{j+1} \sum_{k=0}^{j} \binom{j}{k} (-1)^{j-k} s^k$$

$$= \sum_{k=0}^{n-1} s^k \sum_{j=k}^{n-1} (-1)^{j-k} \binom{j}{k} S_{j+1}.$$

Thus we have proved that

$$P(X > k) = \sum_{j=k}^{n-1} (-1)^{j-k} \binom{j}{k} S_{j+1}. \tag{5}$$

From the relations above we also get

$$\sum_{k=0}^{n-1} (1 + t)^k P(X > k) = \frac{H_X(t) - 1}{t} = \sum_{j=1}^{n} S_j\, t^{j-1}.$$

This gives the following inversion of (5):

$$S_{j+1} = \sum_{k=j}^{n-1} \binom{k}{j} P(X > k).$$

We finish the section with a problem for the reader. Consider three urns, each containing five white or black balls. Suppose urn 1 has two white balls, urn 2 has three white balls and urn 3 has four white balls. Draw one ball from each of the urns. Show that the probability function for the obtained number of white balls is given by $p_0 = 6/125$, $p_1 = 37/125$, $p_2 = 58/125$ and $p_3 = 24/125$.

3.6 Combinations of events II

In Section 2.4 we considered formulas for determining probabilities of certain combinations of events $A_1, \ldots, A_n$. Proofs of these formulas can now be given using zero–one rv's.

(a) *Probability of exactly k events*
Let $p_n(k)$ be the probability that exactly k of n events $A_1, \ldots, A_n$ occur. We asserted in (3) of Section 2.4 that

$$p_n(k) = \sum_{i=k}^{n} (-1)^{i-k} \binom{i}{k} S_i, \tag{1}$$

where

$$S_1 = \sum_i P(A_i); \quad S_2 = \sum_{i<j} P(A_i A_j) \tag{2}$$

and so on. We shall prove this result.

Let $U_1, \ldots, U_n$ be zero–one rv's such that $U_i = 1$ if A_i occurs and $U_i = 0$ otherwise. The sum

$$X = U_1 + \cdots + U_n$$

counts the number of A_i's that occur. Therefore the probability $p_n(k)$ can be written

$$p_n(k) = P(X = k).$$

Using (4) in the preceding section we find

$$p_n(k) = \sum_{j=k}^{n} (-1)^{j-k} \binom{j}{k} S_j, \tag{3}$$

where by (2) in the same section

$$S_j = \sum_{i_1 < \cdots < i_j} E(U_{i_1} \ldots U_{i_j}). \tag{4}$$

More explicitly, we have

$$S_1 = \sum_i E(U_i); \quad S_2 = \sum_{i<j} E(U_i U_j); \quad S_3 = \sum_{i<j<k} E(U_i U_j U_k)$$

and so on. We have now to replace the expectations by probabilities. Since the U's are zero–one rv's, we have for $i < j < k$

$$E(U_i) = P(A_i); \quad E(U_i U_j) = P(A_i A_j); \quad E(U_i U_j U_k) = P(A_i A_j A_k)$$

and so on. Inserting these values in (3) and (4) we obtain (1), and the proof is finished.

(b) *Probability of at least k events*

The probability $P_n(k)$ that at least k of n events $A_1, \ldots, A_n$ occur is given by $P(X > k - 1)$. By (5) in the preceding section and the substitution $i = j + 1$, this can be written

$$P_n(k) = \sum_{i=k}^{n} (-1)^{i-k} \binom{i-1}{k-1} S_i, \tag{5}$$

which proves (4) in Section 2.4.

4

Topics from early days I

In the present chapter and in Chapter 5, we go back to the days of Cardano, Fermat, Pascal, Huygens, the Bernoullis, de Moivre, Montmort and Laplace when probability theory was born and developed into a science. By writing this book, we pay homage in a modest way to these great men. For the historical material we were happy to have had at our disposal the excellent books by Todhunter (1865) and Hald (1990). Another interesting work on the history of probability is Maistrov (1974). Stigler (1986) is an important book on the history of statistics and certain parts of probability. The elementary textbook by Blom (1989b) contains some sections with classical problems of probability.

4.1   Cardano – a pioneer

Gerolamo Cardano (1501–1576) is remembered by mathematicians for his contributions to the solution of the cubic and biquadratic equations. However, he was also a pioneer in probability. His main work in probability, *Liber de Ludo Aleae* (The Book on Games of Chance), was unfortunately not published until 1663, more than a hundred years after it was written. If this book, containing only fifteen folio pages, had been known by the public, say, a century earlier, it would have been the first treatise on our science, which undoubtedly would have honoured Cardano as 'father of probability'.

Cardano's book is very entertaining and written in a lively style. It can be best described as a gambler's manual. Apart from probabilistic discussions it contains rules for different games and advice on how to avoid cheating. The content is easy to grasp for the most part, but there are some unintelligible passages. His advice to players is worth comtemplating even today: 'Play is a very good test of a man's patience or impatience. The greatest advantage in gambling comes from not playing at all. But there is very great utility in it as a test of patience, for a good man will refrain from anger even at the moment of rising from the game in defeat.'

Although *Rabbi Ben Ezra* seems to have had some idea about probability when studying combinations and permutations in connection with astrology, Cardano was the first scholar who mastered some fundamental probabilistic rules. In his book, he often uses the classical definition of probability (or the equivalent notion of odds) formulated today as follows:

A probability p is obtained by dividing the number f of favourable cases by the number c of possible cases. Cardano also knew some combinatorics; see Hald (1990, p. 40).

When considering throws with a symmetric die, Cardano says that the 'circuit' is six, meaning that there are six equally likely cases. For throws with three dice 'the circuit consists of $6 \times 6 \times 6 = 216$ cases'. He correctly counts the number of favourable cases for many problems and calculates the fraction $p = f/c$, or the corresponding odds. However, his arguments are not always crystal clear. For example, when talking about the throw of one die, he says: 'One-half of the total number of faces always represents equality; thus the chances are equal that a given point will turn up in three throws, for the total circuit is completed in six, or again that one of three given points will turn up in one throw.'

His general law of wagers is quite modern: 'So there is one general rule, namely, that we should consider the whole circuit and the number of those casts which represents how many ways the favourable result can occur and compare that number to the remainder of the circuit, and, according to that proportion, should the mutual wagers be laid so that one may contend on equal terms'.

Finally, it should be mentioned that Cardano masters the multiplication formula p^n for n occurrences of an event when an experiment is repeated n times independently, or at least something very close to this fundamental rule. As one of several applications, he considers three independent repetitions of throws, each throw being made with three dice. He finds that the probability that at least one ace shall be obtained in one throw is 91/216, and infers that the probability that this event will occur in each of the three repetitions is equal to $(91/216)^3 = 753{,}571/10{,}077{,}696 \approx 0.0748$. This is a very clever result for its time and shows that Cardano deserves a prominent place in the history of probability.

Reference: Ore (1953).

4.2   Birth of probability

Pierre de Fermat (1601–1665) and *Blaise Pascal* (1623–1662) can be regarded as the fathers of what is nowadays called *combinatorial probability*. As mentioned in the preceding section, Cardano was active in this field a hundred years earlier, but Fermat and Pascal were the first undisputed experts on the combinatorial part of classical probability. 'Godfather' at the birth of probability was *Antoine Gombauld, Chevalier de Méré*, a French nobleman with both practical and theoretical interest in gambling. De Méré consulted the young Pascal, who in turn wrote to Fermat. Thus began in the 1650's Pascal's and Fermat's famous correspondence, which has exerted a profound influence upon the development of probability theory.

The names of Fermat and Pascal are also associated with pure mathematics. Fermat is known for 'Fermat's last theorem', the unproven statement that the equation

$$x^n + y^n = z^n$$

has no integer solution (x, y, z) for $n \geq 3$. Pascal is remembered for 'Pascal's triangle', a triangular arrangement of binomial coefficients. Less known, but vastly more important, is that Pascal introduced mathematical induction and applied this technique when proving theorems concerning binomial coefficients.

We shall now take a simple and well-known example showing the elegant type of reasoning used in their correspondence by quoting from a letter from Pascal to Fermat in 1654. (The translation has been taken from Hald (1990, p. 54).)

'If one undertakes to throw a *six* with one die, the advantage of undertaking it in 4 throws is as 671 to 625. If one undertakes to throw a *double-six* with two dice, there is a disadvantage of undertaking it in 24 throws. And nevertheless 24 is to 36 (which is the number of faces of two dice) as 4 to 6 (which is the number of faces of one die).'

Let us describe this in more familiar terms:

(a) *One die*

In one throw, the probability of not getting 6 is 5/6. In four throws, the probability of not getting 6 is $(5/6)^4 = 625/1{,}296$. Hence the probability of obtaining at least one 6 in 4 throws is $1 - 625/1{,}296 = 671/1{,}296 > 1/2$.

(b) *Two dice*

In one throw with two dice, the probability of not getting a double-six is 35/36. In 24 double throws, the probability of not getting any double-six is $(35/36)^{24}$ and hence the probability of getting at least one double-six is $1 - (35/36)^{24} < 1/2$.

Another problem discussed by Pascal and Fermat is the following: Three players A, B and C participate in a game consisting of independent rounds. Each player wins a round with probability 1/3. The game is over when A has won one round or either B or C has won two rounds. Find the winning probabilities of each player.

A more difficult problem considered by Pascal and Fermat in their correspondence is the division problem to be discussed in the next section.

The correspondence between Pascal and Fermat inspired the Hungarian probabilist *Alfred Rényi* to publish in 1969 the beautiful little book *Briefe über die Wahrscheinlichkeit* (Letters on Probability).

4.3　The division problem

Two persons A and B participate in a game. In each round, A wins with probability p and B with probability $q = 1 - p$. The game ends when one of the players has won r rounds; he is the winner. The winner receives an amount 1, which we call the stake. For some reason, the play stops in advance. At this moment, A has won $r - a$ rounds and B has won $r - b$ rounds. How should the stake be divided between the two players?

This is the *division problem*, also called the *problem of points*; it occupies an important place in the history of probability. The problem was discussed by Cardano in the 16th century, but his solution was incorrect. It was one of the problems proposed to Pascal by de Méré, and it was analysed in 1654 in the famous letters exchanged by Fermat and Pascal; see the preceding section.

Pascal introduced the fruitful idea that the stake should be divided between the players according to their winning probabilities if the game were played to the end. Based on this approach, Pascal and Fermat provided the first correct solution, assuming that the players win a round with the same probability $\frac{1}{2}$. The general solution, for players of unequal skill, was given by *John Bernoulli* (1667–1748), a brother of the famous James Bernoulli.

We shall give two solutions, using modern notation. First observe that the game must terminate at, or before, round no. $2r - 1$, for then certainly one player has won r rounds. It therefore suffices to consider what would happen if $(2r - 1) - (r - a) - (r - b) = a + b - 1$ further rounds are played. We discriminate between two cases:

1. Of the $a + b - 1$ additional rounds, A wins at least a rounds and hence B less than b rounds. Clearly, A then wins the game.

2. B wins at least b rounds and hence A less than a rounds. Player B then wins the game.

Therefore, it seems correct to give A and B the parts $p(a, b)$ and $q(a, b)$ of the stake, respectively, where $p(a, b)$ is the probability that A wins at least a of the $a + b - 1$ rounds and $q(a, b)$ is the probability that B wins at least b of the games. Clearly, these two probabilities have the sum 1.

(a) *First solution*

According to the binomial distribution, the probability that A wins i of n rounds is equal to

$$\binom{n}{i} p^i q^{n-i}.$$

Hence we obtain

$$p(a, b) = \sum_{i=a}^{a+b-1} \binom{a + b - 1}{i} p^i q^{a+b-1-i}. \tag{1}$$

(b) *Second solution*

It is seen that A wins a of the $a + b - 1$ games if he wins $a - 1$ of the first $a - 1 + i$ rounds and also the $(a + i)$th round. Here i is an integer which can assume the values $0, 1, \ldots, b-1$. Hence by the negative binomial distribution (see Section 7.3)

$$p(a, b) = \sum_{i=0}^{b-1} \binom{a - 1 + i}{i} p^a q^i. \tag{2}$$

Example 1

Take $p = 2/5$, $q = 3/5$, $r = 15$. Further, suppose that the players agree to stop the game after round 10 and that A and B have then won 3 and 7 rounds, respectively. How should the stake be divided?

We have $a = 12$, $b = 8$. Formula (1) shows that

$$p(12, 8) = \sum_{i=12}^{19} \binom{19}{i} \left(\frac{2}{5}\right)^i \left(\frac{3}{5}\right)^{19-i} \approx 0.0352.$$

Hence A should receive approximately 3.5 % of the stake and B, 96.5 %.

Example 2

If the players have equal skill and $a = 2$, $b = 3$, player A should receive 11/16 of the stake. We leave the calculations to the reader.

4.4 Huygens's second problem

The Dutchman *Christiaan Huygens* (1629–1695) is a great name in the history of probability. Among the many problems discussed in his important book *De Ratiociniis in Ludo Aleae* (On Reasoning in Games of Chance) we shall discuss two and begin with an easy problem often called *Huygens's second problem*; the other problem is discussed in the next section. Our version of the second problem is more general than Huygens's.

Three players A, B and C participate in the following game. A bowl contains a white and b black balls. They draw a ball with replacement in the order $ABCABC\ldots$ until one player obtains a white ball; he is the winner. Find the winning chances of the players.

Let p, q, r be the probability that A, B, C, respectively, wins the game. Set $\alpha = a/(a+b)$, $\beta = b/(a+b)$. Two cases may occur at the first drawing: (i) If A obtains a white ball, he is the winner, which occurs with probability α; (ii) if A obtains a black ball, he becomes the last of the three in the following sequence of drawings with participants $BCABCA\ldots$; hence in this case he wins with probability r. Also, B becomes the first and C the second. This gives the relations

$$p = \alpha + \beta r; \quad q = \beta p; \quad r = \beta q.$$

Solving this system of equations with respect to p, q, r we find

$$p = \frac{\alpha}{1 - \beta^3}; \quad q = \frac{\alpha\beta}{1 - \beta^3}; \quad r = \frac{\alpha\beta^2}{1 - \beta^3}.$$

The interested reader may perhaps generalize the game to n participants $A_1, A_2, \ldots, A_n$. Defining α and β as before and denoting the winning probability of A_i by p_i, show that

$$p_i = \frac{\alpha\beta^{i-1}}{1 - \beta^n}$$

for $i = 1, 2, \ldots, n$.

4.5   Huygens's fifth problem

We assume in this section that the reader is familiar with linear difference equations.

In the correspondence between Pascal and Fermat in the 1650s, the following problem was discussed. Players A and B have 12 counters each. They play in each round with three dice until either 11 points or 14 points are obtained. In the first case, A wins the round and gives a counter to B; in the latter case, B wins and gives a counter to A. The game goes on until one player obtains all the counters; he is the winner. Find the probability that A wins the game. This is called *Huygens's fifth problem*. It was given with an answer but without a solution in Huygens's book mentioned in the preceding section. The problem was solved by James Bernoulli (see Section 4.7), but his solution was not published until 1713, eight years after his death.

We shall solve a more general problem. Assume that, initially, A has a counters and B has b counters and that they win a round with probability p and $q = 1 - p$, respectively. We assume that $p \neq q$.

Let $P(a, b)$ and $Q(a, b)$ be the probabilities that A and B, respectively, *lose* all counters. We call this *the ruin problem for players of unequal skill*. It was first solved by de Moivre; the first solution based on a difference equation was given by the Dutch scholar *Nicolaas Struyck* (1687–1769), which was a major achievement. (For the ruin problem for players of equal skill see Section 1.5.)

For the solution we use Struyck's method. After the first round, A has either $a+1$ or $a-1$ counters. Hence we obtain the difference equation

$$P(a,b) = pP(a+1,b-1) + qP(a-1,b+1).$$

There are two boundary conditions, $P(0, a+b) = 1$ and $P(a+b, 0) = 0$. The solution is [compare Feller (1968, p. 344)]

$$P(a,b) = \frac{(q/p)^{a+b} - (q/p)^a}{(q/p)^{a+b} - 1}.$$

Interchanging p and q, and a and b, we further obtain

$$Q(a,b) = \frac{(p/q)^{a+b} - (p/q)^b}{(p/q)^{a+b} - 1}.$$

It is found that these expressions add up to 1, which implies that an unending game is impossible.

When $a = b$, we find after a reduction the following quotient between A's and B's probabilities of losing the game:

$$\frac{P(a,a)}{Q(a,a)} = \left(\frac{q}{p}\right)^a.$$

We now return to Huygens's fifth problem. We then have $a = 12$, and a simple calculation shows that $p = 9/14$ and $q = 5/14$. Hence

$$\frac{P(12,12)}{Q(12,12)} = \left(\frac{5}{9}\right)^{12} \approx 0.0009.$$

A's chance of *winning* all counters divided by B's chance is $(9/5)^{12} \approx 1,157$.

4.6 Points when throwing several dice

A symmetrical device, here called a die, has m faces. The ith face shows i points, $i = 1, 2, \ldots, m$. We throw n such dice simultaneously. Find the probability function of the total number of points obtained.

This problem has been discussed by many of the old probabilists. In *Liber de Ludo Aleae*, Cardano enumerated favourable cases, as did *Galileo Galilei* (1564–1642) for $n = 3$. It was also treated by de Moivre and by *Pierre Rémond de Montmort* (1678–1719), who gave a combinatorial solution, using inclusion and exclusion; compare Section 2.4.

De Moivre solved the problem using generating functions. We will show his solution, using modern notation.

Let Z be the total number of points and set

$$Z = X_1 + X_2 + \cdots + X_n,$$

where X_i is the number of points shown on the ith die. The probability generating function of each X_i is given by

$$G_{X_i}(s) = E(s^{X_i}) = \frac{1}{m}(s + s^2 + \cdots + s^m).$$

Since the X's are independent, the probability generating function of Z becomes (compare Section 3.2)

$$G_Z(s) = E(s^{\Sigma X_i}) = \prod_{i=1}^{n} E(s^{X_i}) = \frac{1}{m^n}(s + s^2 + \cdots + s^m)^n.$$

The probability function $P(Z = k)$ is obtained by collecting the coefficients of s^k in the expansion of $G_Z(s)$. Write

$$(s + s^2 + \cdots + s^m)^n = s^n(1 - s^m)^n(1 - s)^{-n}$$

$$= s^n \sum_{i=0}^{n} (-1)^i \binom{n}{i} s^{im} \sum_{j=0}^{\infty} \binom{-n}{j}(-s)^j$$

$$= s^n \sum_{i=0}^{n} (-1)^i \binom{n}{i} s^{im} \sum_{j=0}^{\infty} \binom{n+j-1}{j} s^j.$$

Set $n + im + j = k$ for $i = 0, 1, \ldots, [(k-n)/m]$, where $[\ ]$ denotes the integer part. Replacing j by $k - n - im$ we obtain

$$\binom{n+j-1}{j} = \binom{k - im - 1}{k - im - n} = \binom{k - im - 1}{n - 1}.$$

The final result is

$$P(Z = k) = \sum_{i=0}^{[(k-n)/m]} (-1)^i \binom{n}{i} \binom{k - im - 1}{n - 1} \frac{1}{m^n}. \qquad (1)$$

It is impressive that de Moivre was able to solve the problem in this way. He constructed a new tool for probability calculations, which later, in the hands of Laplace, proved very important for the development of the new science.

Example

Ten fair dice are thrown at random. What is the chance P that the dice together show 27 points?

The probability P is given by taking $k = 27$, $n = 10$, $m = 6$ in (1). But let us illustrate the derivation of (1) with this explicit example. Imagine that the dice are numbered $1, 2, \ldots, 10$ and that the points are $X_1, X_2, \ldots, X_{10}$. Here the X's are independent rv's taking the values $1, 2, \ldots, 6$ with the same probability. By a total enumeration of all $6^{10} = 60{,}466{,}176$ cases the probability $P(X_1 + X_2 + \cdots + X_{10} = 27)$ could in principle be computed, but this is very impractical. The approach with generating functions works as follows. First we have for each X_i

$$G_{X_i}(s) = \frac{1}{6}(s + s^2 + \cdots + s^6) = \frac{s(1 - s^6)}{6(1 - s)}.$$

Thus, the probability asked for is given by the coefficient of s^{27} in

$$G_{X_1 + \cdots + X_{10}}(s) = \left[\frac{s(1 - s^6)}{6(1 - s)} \right]^{10}.$$

The answer is

$$P = \frac{1}{6^{10}} \left[-\binom{10}{0}\binom{-10}{17} + \binom{10}{1}\binom{-10}{11} - \binom{10}{2}\binom{-10}{5} \right]$$

$$= \frac{1}{6^{10}} \left[\binom{26}{9} - 10\binom{20}{9} + 45\binom{14}{9} \right] = \frac{2{,}665}{104{,}976} \approx 0.0254.$$

4.7   Bernoulli and the game of tennis

James Bernoulli (1654–1705) is known by all students of probability for Bernoulli's theorem, which states that the relative frequency of an event in a sequence of independent trials converges in probability to the probability of the event. He is also known for the Bernoulli numbers, which are related to the Euler numbers and the Stirling numbers; see Chapters 5 and 6. James Bernoulli's main work is *Ars Conjectandi* (see Section 5.1). His *Letter to a Friend on the Game of Tennis* is much less known. It contains solutions of some problems connected with what is nowadays called random walk; see Chapter 10. In our presentation of one of his problems, we follow the exposition given by Hald (1990, p. 241).

Consider a game of tennis played according to the usual rules. Instead of the standard scores 0, 15, 30, 40, game, we use points $0, 1, \ldots$. The game then stops at 4–0, 4–1, 4–2, 5–3, 6–4, 7–5, etc. or 0–4, etc.; see the boundaries in Figure 1.

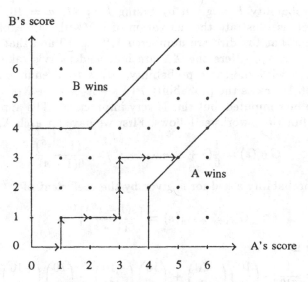

Fig. 1. Stopping region at tennis.

Let us remark in passing that the figure reminds us of modern sequential analysis introduced by *Abraham Wald*. Tennis can be regarded as a sequential procedure, and many lecturers have used this game for illustrating to their students the fruitful idea: stop at boundaries which depend on the successive results.

Bernoulli introduces the following model for a game of tennis. Players A and B win a point with probability p and $q = 1 - p$, respectively. The successive points are assumed to be won independently (an assumption which severely limits the practical applicability of the model, as also does the assumption of a constant winning probability p).

Let $P(i, j)$ be the probability that A wins the game, given that A and B have won i and j points, respectively. One of the problems stated by Bernoulli was to find $P(i, j)$ for all possible values of i and j and for values of p such that $p/q = n$, where n is an integer. Today this is an elementary problem, but was not three hundred years ago.

Bernoulli's solution is elegant: He starts from the recurrence relation

$$P(i, j) = pP(i + 1, j) + qP(i, j + 1).$$

By a two-fold application of this formula he infers that

$$P(3, 3) = p^2 P(5, 3) + 2pq P(4, 4) + q^2 P(3, 5).$$

Since $P(5,3) = 1$, $P(3,5) = 0$, $P(4,4) = P(3,3)$, he obtains the first answer:

$$P(3,3) = \frac{p^2}{p^2 + q^2} = \frac{n^2}{n^2 + 1}.$$

By similar calculations he determines all quantities $P(i,j)$ when $p/q = n$ is 1, 2, 3, 4. The most interesting quantity $P(0,0)$ is given by

$$P(0,0) = \frac{n^7 + 5n^6 + 11n^5 + 15n^4}{n^7 + 5n^6 + 11n^5 + 15n^4 + 15n^3 + 11n^2 + 5n + 1}.$$

Using this relation, we find

n	1	2	3	4
$P(0,0)$	$\frac{1}{2}$	$\frac{208}{243}$	$\frac{243}{256}$	$\frac{51,968}{53,125}$

We end the section with a problem for the reader. One of Bernoulli's results is the following: If $p/q = 2$, then $P(0,2) = 208/405 \approx 0.5136$. Hence if the weaker player B is given a handicap of two points, the game is almost fair. Show this.

5

Topics from early days II

In this chapter, we present more problems and snapshots from the early days of probability. For references, see the introduction to Chapter 4.

5.1 History of some common distributions

We nowadays use several standard discrete distributions without always paying attention to their history. In this section, we give a few glimpses from the history of the binomial, multinomial, hypergeometric and Poisson distributions. A good general reference to these distributions is Johnson and Kotz (1969).

(a) *Binomial distribution*

An rv X has a *binomial distribution* with parameters n and p if the probability function is given by

$$P(X = k) = \binom{n}{k} p^k q^{n-k}, \tag{1}$$

where $k = 0, 1, \ldots, n$; $0 \leq p \leq 1$ and $q = 1 - p$. As remarked in 'Symbols and formulas' at the beginning of the book, we use the code name $X \sim \text{Bin}(n, p)$.

As is well known, the binomial distribution occurs in certain common types of random trials. Assume that the probability of the occurrence of an event A in a single trial is p. If n independent trials are performed and X is the frequency of A, then $X \sim \text{Bin}(n, p)$.

This important property of the binomial distribution was known to James Bernoulli, who discusses the distribution in his famous treatise *Ars Conjectandi* (The Art of Conjecturing) published in 1713, eight years after his death. Bernoulli also considered the probability P_{mn} that the event A occurs at least m times. Let Q_{mn} be the probability of the complementary event that A occurs at most $m - 1$ times. Bernoulli used the recursion

$$Q_{mn} = pQ_{m-1,n-1} + qQ_{m,n-1}.$$

By induction he managed to prove that

$$Q_{mn} = \sum_{k=0}^{m-1} \binom{n}{k} p^k q^{n-k}.$$

The binomial distribution was also considered by the French nobleman Montmort in the second edition of his book *Essay d'Analyse sur les Jeux de Hazard* (Analysis of Games of Chance) published in 1713; the first edition was published in 1708. Already in 1654, Pascal derived the distribution for the case $p = \frac{1}{2}$, but it is not entirely clear whether he regarded it as a probability distribution or as a set of expectations showing how much to stake on the results $X = 0, 1, \ldots, n$.

It should be remembered that the terms in (1) can be obtained from the expansion of

$$(p + q)^n.$$

The history of the binomial distribution is therefore linked to the antecedents of the binomial expansion, although it took a long time before the probabilistic content of this expansion was recognized.

(b) *Multinomial distribution*

The rv's $X_1, \ldots, X_r$ have a *multinomial distribution* if their probability function is given by

$$P(X_1 = k_1, \ldots, X_r = k_r) = \frac{n!}{k_1! \cdots k_r!} p_1^{k_1} \cdots p_r^{k_r}, \tag{2}$$

where $\sum p_i = 1$ and $\sum k_i = n$.

The multinomial distribution occurs in independent random trials. Suppose that each trial has r different outcomes occurring with probabilities $p_1, \ldots, p_r$. Perform n such trials independently and count the frequencies $X_1, \ldots, X_r$ of the different outcomes. These frequencies have a multinomial distribution.

The multinomial distribution was discussed by Montmort in the first edition of his book mentioned above. The case of general chances was first considered in 1710 by de Moivre, in the remarkable paper *De Mensura Sortis* (On the Measurement of Chance) published two years later in the *Philosophical Transactions*.

(c) *Hypergeometric distribution*

If the rv X has the probability function

$$P(X = k) = \binom{a}{k} \binom{b}{n-k} \bigg/ \binom{a+b}{n}, \tag{3}$$

where $0 \le k \le a$, $0 \le n - k \le b$, then X has a *hypergeometric distribution* with parameters n, a and b.

This distribution occurs in connection with urn models. Let us consider an urn with a white and b black balls. If n balls are drawn at random, one at a time without replacement, the number X of white balls obtained has a hypergeometric distribution.

As a generalization, let us consider an urn with balls of r different colours; a_i balls of colour i, $i = 1, \ldots, r$. Draw n balls again without replacement, and let X_i be the number of balls of colour i which is obtained. The rv's $X_1, \ldots, X_r$ then have an *r-variate hypergeometric distribution* with probability function

$$P(X_1 = k_1, \ldots, X_r = k_r) = \binom{a_1}{k_1} \cdots \binom{a_r}{k_r} \Big/ \binom{a_1 + \cdots + a_r}{n}, \quad (4)$$

where $\sum k_i = n$ and $0 \le k_i \le a_i$ for $i = 1, \ldots, r$.

Hypergeometric probabilities were known to the 'fathers of probability'. Huygens examined several problems with probabilistic content in *De Ratiociniis in Ludo Aleae*; compare Section 4.4. One of the many problems it contained, called *Huygens's fourth problem*, is the following: Two players have twelve chips, four white and eight black; A wagers with B that by drawing seven chips blindfolded, he will get three white chips. The question is: what is the ratio of A's chances to B's?

Huygens solved the problem by a recursive method. His problems were later discussed by several probabilists, among them James Bernoulli and Montmort; they showed, using the now familiar combinatorial argument, that A obtains three white chips with probability

$$\binom{4}{3}\binom{8}{4} \Big/ \binom{12}{7} = \frac{35}{99}.$$

Hence B wins with probability $1 - 35/99 = 64/99$, and the ratio sought by Huygens is $35/64 \approx 0.5469$.

(d) *Poisson distribution*

An rv X has a *Poisson distribution* with parameter m if its probability function can be written

$$P(X = j) = \frac{m^j}{j!} e^{-m}, \quad (5)$$

where $j = 0, 1, 2, \ldots$. We write $X \sim \mathrm{Po}(m)$.

The Poisson distribution is often used as an approximation to a binomial distribution $\mathrm{Bin}(n, p)$ when p is small. This is a consequence of the following limiting result:

When n goes to infinity and p goes to zero in such a way that $np \to \lambda$, we have for any given k

$$\binom{n}{k} p^k q^{n-k} \to \frac{\lambda^k}{k!} e^{-\lambda}.$$

This was essentially the way the French mathematician *Siméon Denis Poisson* (1781–1840) derived the Poisson distribution in his book *Recherches sur la probabilité des jugements en matière criminelle et en matière civile, précedées des règles générales du calcul des probabilités*, from 1837. (However, the distribution occurred about a century earlier implicitly in the works of de Moivre.)

Poisson's discovery seems to have been unnoticed for a long time, for there are very few references to it between 1837 and 1898. Bortkewitsch then wrote his important book *Das Gesetz der kleinen Zahlen* (The Law of Small Numbers), which has had a profound influence on the statistical applications of the Poisson distribution.

A good reference to the Poisson distribution is Haight (1967).

5.2   Waldegrave's problem I

Waldegrave's problem was proposed to Montmort by Waldegrave, an English gentleman, who solved it himself in a special case. It was also considered by *Nicholas Bernoulli* (1687–1759), a nephew of James Bernoulli, by de Moivre and by Struyck. For a detailed exposition of various aspects on the problem and on the solutions available, see Hald (1990, p. 378).

The rules leading to the problem are the following: The $r + 1$ players $A_1, \ldots, A_{r+1}$, which are of equal skill, participate in a game consisting of several rounds. They play in the 'circular' order

$$A_1, A_2, \ldots, A_{r+1}, A_1, A_2, \ldots, A_{r+1}, A_1, A_2, \ldots .$$

First A_1 plays with A_2; the loser does not enter the game until all the other people have played; the winner plays the next round with A_3. Similarly, the loser of the second round does not enter the game until all the other people have played; the winner plays against the next gambler in turn, that is, against A_4, and so on until one player succeeds in winning over all his adversaries in immediate succession; he is the winner of the whole game.

We shall derive the probability function of the number N of rounds played when the game stops.

Let p_n be the probability that $N = n$. We have, of course,

$$p_1 = p_2 = \cdots = p_{r-1} = 0. \tag{1}$$

We now consider two cases:

1. $n = r$: This case occurs if the winner of round 1 (that is, A_1 or A_2) also wins the $r - 1$ following rounds. Hence we have

$$p_r = (\tfrac{1}{2})^{r-1}. \tag{2}$$

2. $n \geq r + 1$: The p's satisfy the recurrence relation

$$p_n = (\tfrac{1}{2})p_{n-1} + (\tfrac{1}{2})^2 p_{n-2} + \cdots + (\tfrac{1}{2})^{r-1} p_{n-r+1}. \qquad (3)$$

Using (1), (2) and (3) we may determine all the p's. We shall give a combinatorial proof of (3), which we are very fond of.

Associate with each round, from the second and onwards, a 1 if the winner of the round also won the preceding round, and a 0 otherwise. These two alternatives occur with the same probability $\tfrac{1}{2}$ and the outcomes are independent. The game stops when $r - 1$ 1's in succession have been obtained. (This trick simplifies the thinking, since we may entirely forget the individual players.)

We shall now use the classical definition of probability: we determine p_n by dividing the number of favourable sequences by the number of possible binary sequences of length n. The detailed proof will be given only for five players, that is, for the special case $r = 4$.

Let a_n be number of favourable sequences of length n; they end with $1\,1\,1$ and do not contain $1\,1\,1$ inside. For example, when $n = 6$, there are four such sequences $0\,1\,0\,1\,1\,1, 0\,0\,0\,1\,1\,1, 1\,0\,0\,1\,1\,1, 1\,1\,0\,1\,1\,1$. We have $a_1 = 0, a_2 = 0, a_3 = 1$ and the simple recurrence

$$a_n = a_{n-1} + a_{n-2} + a_{n-3} \qquad (4)$$

for $n = 4, 5, \ldots$. The proof is easy. When $n = 3$ there is only one favourable sequence, $1\,1\,1$, and so $a_3 = 1$. When $n \geq 4$ a favourable sequence of length n *either* begins with 0 followed by a favourable sequence of length $n - 1$, *or* begins with $1\,0$ followed by a favourable sequence of length $n - 2$, *or* begins with $1\,1\,0$ followed by a favourable sequence of length $n - 3$. This leads to (4).

We now divide a_n by the number 2^n of possible sequences. Hence from (4) we infer that

$$p_n = (\tfrac{1}{2})p_{n-1} + (\tfrac{1}{2})^2 p_{n-2} + (\tfrac{1}{2})^3 p_{n-3},$$

in agreement with (3).

The reader will now understand how the proof runs for a general r; the number of a's on the right side of (4) is then $r - 1$.

The special case $r = 3$ is worth mentioning. The recurrence (4) then reduces to an old friend from combinatorics: the Fibonacci sequence with starting values $a_1 = 0, a_2 = 1$ and recurrence $a_n = a_{n-1} + a_{n-2}$ for $n = 3, 4, \ldots$.

It is also possible to determine the winning probabilities of the players; see Nicholas Bernoulli's solution to this problem described in Hald's book (1990, p. 380). See also Section 14.6.

5.3 Petersburg paradox

A person is invited by a rich friend to participate in the following game. A fair coin is tossed until heads appears for the first time. If this happens on the kth toss, the player receives 2^k dollars, where $k = 1, 2, \ldots$. Determine the expectation of the amount that the player receives.

Let Y be this amount. The number X of tosses has the probability function $P(X = k) = (\frac{1}{2})^k$, where $k = 1, 2, \ldots$. We find

$$E(Y) = E(2^X) = 2(\tfrac{1}{2}) + 2^2(\tfrac{1}{2})^2 + \cdots .$$

The sum diverges and hence $E(Y)$ is infinite. The game seems to be exceptionally favourable to the player, since his rich friend apparently loses 'an infinite amount of money'. This paradox deserves a closer examination.

First, suppose that the game is played only once. It is seen that $X \leq 6$ with the probability

$$\frac{1}{2} + \frac{1}{2^2} + \cdots + \frac{1}{2^6} = 1 - \frac{1}{2^6} = \frac{63}{64}.$$

This tells us that there is a large probability that the rich friend escapes with paying at most $2^6 = 64$ dollars. Similarly, the chance is $1{,}023/1{,}024$, that is, almost 1, that the friend has to pay at most $2^{10} = 1{,}024$ dollars. To let the person play once thus seems rather safe for the rich friend.

It is worse for the friend if the player makes use of the offer many times. Then it is no longer possible to disregard the fact that the mean is infinite, for the mean shows, as we know, what will be paid on the average. The wealthy friend may perhaps demand a stake s which makes the game fair. Then $E(Y) - s$ should be zero, and, paradoxically enough, the stake must be 'infinite'.

Such a game is impossible to play, for example, in Monte Carlo, and must be modified in some way. For example, it may be prescribed that the amount 2^k dollars is paid to the player only if the number k of tosses is at most equal to a given number r. Then the expectation of the payment becomes

$$2(\tfrac{1}{2}) + 2^2(\tfrac{1}{2})^2 + \cdots + 2^r(\tfrac{1}{2})^r = r,$$

and it is possible to make the game fair.

Another way of making the game fair is to let the total stake depend on the number of games played. Suppose that a person decides in advance to play the game n times. Feller (1968, p. 251) has proved that a fair stake for each of these games is $\log_2 n$ provided that n is large. Moreover, Martin–Löf (1985) has shown that if this stake is increased to $2^m + \log_2 n$, where 2^m is reasonably large, there is only the small probability $\approx 1.8 \cdot 2^{-m}$ that the rich man cannot cover his expenses. Whether the player is willing to pay this amount is another question.

The Petersburg paradox presumably got its name because the mathematician *Daniel Bernoulli* (1700–1782), a nephew of James Bernoulli, discussed it in proceedings published by the Academy in St. Petersburg. Many other mathematicians and probabilists have considered the problem, for example, D'Alembert, Buffon, Condorcet and Laplace.

A good article on the Petersburg paradox is found in *Encyclopedia of Statistical Sciences* (1982–1988); two more sources are Todhunter (1865) and Dutka (1988).

5.4   Rencontre I

A group of n fans of the winning football team throw their hats high into the air. The hats fall dawn randomly, one hat to each of the n fans. What is the probability that no fan gets his own hat back?

This is a version of what is often called the *rencontre problem*. Another name is the *matching problem*. It can be formulated in various ways; see, for instance, Example 1 in Section 2.6. The problem has a long history going back to the beginning of the 18th century when it was first discussed by Montmort in connection with the card game *le Jeu du Treize* (The Game of Thirteen). Also, de Moivre and the great mathematician *Leonhard Euler* (1707–1783) studied the problem. We will discuss other versions of the rencontre problem in Sections 7.8 and 8.5.

Consider a *random permutation* $(\pi_1, \pi_2, \ldots, \pi_n)$ of $1, 2, \ldots, n$. It is obtained by drawing one of the $n!$ possible permutations of $1, 2, \ldots, n$ at random. If $\pi_k = k$, we say, using a modern term, that k is a *fixed point* of the random permutation. For example, if $n = 9$ and we obtain the permutation

Position	1	2	3	4	5	6	7	8	9
Permutation	1	6	8	2	7	5	4	3	9

then there are the two fixed points: 1 and 9.

The rencontre problem may be stated in the following way: What is the probability that a random permutation has no fixed point?

We shall answer this question by solving a more general problem. Consider the events $A_1, \ldots, A_n$, where A_i is the event 'i is a fixed point'. Let $P_n(k)$ be the probability that at least k of the events A_i occur. Noting that the events are exchangeable, we obtain from formula (2) in Section 2.4

$$P_n(k) = \sum_{i=k}^{n} (-1)^{i-k} \binom{i-1}{i-k} \binom{n}{i} P(A_1 \cdots A_i). \qquad (1)$$

In order to find $P(A_1 \cdots A_i)$, we use the classical rule for determining probabilities: divide the number of favourable cases by the number of possible cases. The total number of cases is $n!$. Favourable cases are those where the integers $1, 2, \ldots, i$ have positions $1, 2, \ldots, i$, respectively, while the remaining $n - i$ integers appear in any order; this gives $(n - i)!$ cases. Hence we have

$$P(A_1 \cdots A_i) = \frac{(n - i)!}{n!}.$$

Inserting this in (1) and reducing, we obtain

$$P_n(k) = \sum_{i=k}^{n} (-1)^{i-k} \binom{i-1}{i-k} \frac{1}{i!}.$$

Taking $k = 1$ we find

$$P_n(1) = \frac{1}{1!} - \frac{1}{2!} + \cdots + (-1)^{n-1} \frac{1}{n!}.$$

This is the probability that there is at least one fixed point. Therefore, the probability that no fixed point occurs is

$$1 - \frac{1}{1!} + \frac{1}{2!} - \cdots + (-1)^n \frac{1}{n!}, \tag{2}$$

and so this is the answer to the hat problem given at the beginning of the section. The sum (2) contains the first $n + 1$ terms in the power series expansion of e^{-1}. Therefore the answer is approximately $e^{-1} \approx 0.3679$. Formula (2) was first given by Montmort.

Here is a problem for the interested reader: Ten boys throw their hats into the air. Show that the probability that at least half of them gets their own hat back equals $829/226,800 \approx 0.0037$.

5.5   Occupancy I

In the present section and in Section 8.7 we shall discuss so called *occupancy problems*. We begin with the *classical occupancy problem*, first studied by de Moivre in his remarkable paper from 1712, *De Mensura Sortis* (On the measurement of chance).

A symmetrical die with r faces is thrown n independent times. What is the probability P_r that all faces appear at least once? Let A_i be the event that the ith face does not appear in n throws. The probability that the ith face does not occur at a single throw is $1 - 1/r$. Because of the independence of the throws, we have

$$P(A_i) = \left(1 - \frac{1}{r}\right)^n.$$

Furthermore, the A_i's are exchangeable events; see Section 2.2 for a definition. Hence we have, for any j integers $i_1 < \cdots < i_j$ chosen among $1, 2, \ldots, r$, that

$$P(A_{i_1} \cdots A_{i_j}) = \left(1 - \frac{j}{r}\right)^n$$

is the probability that none of the faces $i_1, \ldots, i_j$ appear at n throws. By the inclusion–exclusion formula (see Section 2.4) we have

$$P(A_1 \cup \cdots \cup A_r) = S_1 - S_2 + \cdots + (-1)^{r-1}S_r,$$

where

$$S_j = \sum_{i_1 < \cdots < i_j} P(A_{i_1} \cdots A_{i_j}).$$

In the present case, the general expression for S_j reduces to

$$S_j = \binom{r}{j}\left(1 - \frac{j}{r}\right)^n$$

and so we obtain

$$P(A_1 \cup \cdots \cup A_r) = \sum_{j=1}^{r}(-1)^{j-1}\binom{r}{j}\left(1 - \frac{j}{r}\right)^n.$$

This is the probability that at least one of the faces does not appear. Taking the complement we find the probability we seek:

$$P_r = \sum_{j=0}^{r}(-1)^{j}\binom{r}{j}\left(1 - \frac{j}{r}\right)^n.$$

Example. A die game

You are offered to bet 15 dollars to participate in the following game: A standard die is thrown 20 times. If any of the six faces does not appear you win 100 dollars, otherwise you lose your stake. Should you participate?

In this game, we may use the results above. We have $r = 6, n = 20$, and the probability to win the game is $1 - P_6 \approx 0.1520$. Hence, the expected profit is approximately $100 \cdot 0.1520 - 15 = 0.20 > 0$. Consequently, in the long run it is advantageous to participate in the game.

(a) *Expected number of missing faces*

Let X be the number of faces which do not occur at n throws. Write

$$X = U_1 + \cdots + U_r$$

as a sum of zero–one rv's: set $U_i = 1$ if the ith face does not occur at any of the n throws and $U_i = 0$ otherwise. Since

$$P(U_i = 1) = P(A_i) = \left(1 - \frac{1}{r}\right)^n,$$

we have

$$E(U_i) = \left(1 - \frac{1}{r}\right)^n$$

and so

$$E(X) = r\left(1 - \frac{1}{r}\right)^n.$$

This is a simple and nice result.

Example. *Four-digit random numbers*

Select 500 four-digit random numbers among $0000, 0001, \ldots, 9999$. The mean number of missing numbers among the 10^4 possible is

$$10,000 \left(1 - \frac{1}{10,000}\right)^{500} \approx 9,512.3.$$

Hence the number of different numbers appearing in the sample is, on the average, equal to

$$10,000 \left[1 - \left(1 - \frac{1}{10,000}\right)^{500}\right] \approx 487.7,$$

and the mean number of numbers appearing at least twice is approximately

$$500 - 487.7 = 12.3.$$

(b) *More about missing faces*

In (a) we determined the expectation $E(X)$ of the number of missing faces. More generally, using (3) in Section 3.5, we find that the jth binomial moment of X is given by

$$E\left[\binom{X}{j}\right] = \sum_{1 \le i_1 < \cdots < i_j \le r} E(U_{i_1} \cdots U_{i_j}) = \binom{r}{j}\left(1 - \frac{j}{r}\right)^n.$$

Also, the probability function of X can be found, using the results of Section 3.5, Subsection (c). The resulting function is

$$P(X = k) = \sum_{j=k}^{r} (-1)^{j-k} \binom{j}{k}\binom{r}{j}\left(1 - \frac{j}{r}\right)^n. \tag{1}$$

Finally, let $Y = r - X$ be the number of different faces which have occurred
at least once. It can be shown by rewriting (1) that the probability function
of Y is given by

$$P(Y = k) = \binom{r}{k}\left(\frac{k}{r}\right)^n q_{kn},\qquad (2)$$

where q_{kn} is the probability that all faces are obtained when throwing a
die with k faces n times; of course, $q_{kn} = 0$ for $k > n$. Taking $k = 0$ and
$r = k$ in (1) we have

$$q_{kn} = \sum_{j=0}^{k}(-1)^j \binom{k}{j}\left(1 - \frac{j}{k}\right)^n.\qquad (3)$$

5.6   Stirling numbers of the second kind

This is a section primarily intended for mathematically minded readers. It
is devoted to *Stirling numbers of the second kind*. They were introduced by
the English mathematician *James Stirling* (1692–1770). More details are
found in Graham, Knuth and Patashnik (1989) and Knuth (1992).

(a) *Definition*

The Stirling numbers of the second kind, here denoted by $\{{n \atop k}\}$, can be
defined as follows (other definitions occur).

   The Stirling number $\{{n \atop k}\}$ shows the number of ways in which a set of n
different objects can be partitioned into k nonempty subsets. For example,
there are seven ways to split a four-element set into two parts:

$$\{1,2,3\}\cup\{4\};\quad \{1,2,4\}\cup\{3\};\quad \{1,3,4\}\cup\{2\};\quad \{2,3,4\}\cup\{1\};$$

$$\{1,2\}\cup\{3,4\};\quad \{1,3\}\cup\{2,4\};\quad \{1,4\}\cup\{2,3\};$$

thus $\{{4 \atop 2}\} = 7$.

   The numbers $\{{n \atop k}\}$ satisfy the recursive relations

$$\left\{{n \atop k}\right\} = \left\{{n-1 \atop k-1}\right\} + k\left\{{n-1 \atop k}\right\},$$

with $\{{n \atop 1}\} = 1$ and $\{{n \atop k}\} = 0$ for $k > n$.

Table 1. Stirling numbers of the second kind $\{{n \atop k}\}$.

$n\backslash k$	1	2	3	4	5	6	7	8
1	1							
2	1	1						
3	1	3	1					
4	1	7	6	1				
5	1	15	25	10	1			
6	1	31	90	65	15	1		
7	1	63	301	350	140	21	1	
8	1	127	966	1,701	1,050	266	28	1

(b) *Relation to the classical occupancy problem*

Perform n trials with k equally likely outcomes. Conceptually, we may obtain the result of a trial in two steps:

1. Partition the sequence of n into k nonempty subsequences (the number of ways is $\{{n \atop k}\}$).
2. Assign outcomes $1, 2, \ldots, k$ to the subsequences, one to each subsequence (the number of ways is $k!$).

As a result, the probability q_{kn} that in n independent trials with k equally likely outcomes all k outcomes occur can be written

$$q_{kn} = \frac{k!\{{n \atop k}\}}{k^n}. \tag{1}$$

This probability is also given by formula (3) in Section 5.5. From there we get the explicit expression

$$\left\{ {n \atop k} \right\} = \frac{k^n q_{kn}}{k!} = \frac{1}{k!} \sum_{j=0}^{k} (-1)^j \binom{k}{j} (k-j)^n.$$

Note that for $k > n$ we have $\{{n \atop k}\} = 0$ and $q_{kn} = 0$.

(c) *Another application*

Let Y be the number of different faces which have occurred in n throws of a symmetric die with r faces. Then, for $k = 1, 2, \ldots$, we have (see (2) of Section 5.5)

$$P(Y = k) = \binom{r}{k} \left(\frac{k}{r} \right)^n q_{kn},$$

with q_{kn} as above; note that $P(Y = k) = 0$ for $k > \min(n, r)$. Summing this probability function and using (1), we obtain the identity

$$\frac{1}{r^n} \sum_{k=1}^{n} \left\{ {n \atop k} \right\} r(r-1) \cdots (r-k+1) = 1,$$

as $r(r-1)\cdots(r-k+1) = 0$ for $k > r$. Hence for all positive integers r,

$$r^n = \sum_{k=1}^{n} \left\{ {n \atop k} \right\} r(r-1)\cdots(r-k+1).$$

As both sides are polynomials in r we have for *any* quantity s

$$s^n = \sum_{k=1}^{n} \left\{ {n \atop k} \right\} s(s-1)\cdots(s-k+1),$$

expressing ordinary powers in terms of descending factorials. With s replaced by $-s$ we obtain

$$s^n = \sum_{k=1}^{n} (-1)^{n-k} \left\{ {n \atop k} \right\} s(s+1)\cdots(s+k-1).$$

These formulas imply the relations

$$E(X^n) = \sum_{k=1}^{n} \left\{ {n \atop k} \right\} E[X(X-1)\cdots(X-k+1)],$$

$$E(X^n) = \sum_{k=1}^{n} (-1)^{n-k} \left\{ {n \atop k} \right\} E[X(X+1)\cdots(X+k-1)].$$

Here is an easy problem: Table 1 indicates that $\left\{ {n \atop 2} \right\} = 2^{n-1} - 1$. Prove this.

5.7 Bayes's theorem and Law of Succession

We shall discuss Bayes's theorem and apply it to Laplace's famous Law of Succession.

(a) *Bayes's theorem*

Let $H_1, H_2, \ldots, H_n$ be events in a probability space Ω. We assume that they are mutually exclusive, have positive probabilities and fill the space completely. Furthermore, let A be an event in which we are specially interested. *Bayes's theorem* states that

$$P(H_i|A) = \frac{P(H_i)P(A|H_i)}{\sum_{j=1}^{n} P(H_j)P(A|H_j)}. \tag{1}$$

Bayes's formula (1) is known to all students of probability. It shows how the conditional probability of H_i given that A has occurred can be found provided that the *prior probabilities* $P(H_i)$ are known. The formula is ascribed to the English clergyman *Thomas Bayes* (c. 1702–1761) which is not entirely correct. Bayes does not mention the formula explicitly in his writings, but examines questions related to it.

According to a personal communication to the authors from A. Hald, Laplace was first to give formula (1); it therefore ought to be called Laplace's formula.

Bayes's theorem is nowadays of fundamental importance for all friends of subjective probability. For objectivists it is mainly a theorem in mathematical probability with valuable but limited application.

The crux of the matter is that the prior probabilities $P(H_j)$ are not often known, except by subjectivists, who assess them subjectively. If they can be regarded as equal, which is sometimes possible, one gets rid of them, since formula (1) then simplifies to

$$P(H_i|A) = \frac{P(A|H_i)}{\sum_{j=1}^n P(A|H_j)}. \tag{2}$$

(b) *Inverse probability*

We shall apply Bayes's theorem to a problem stated by Bayes in his famous *An Essay Toward Solving a Problem in the Doctrine of Chances* printed after his death; see Bayes (1764). Let us quote from the essay:

'Given the number of times in which an unknown event has happened and failed: Required the chance that the probability of its happening in a single trial lies somewhere between any two degrees of probability that can be named.'

Let us consider the following related problem, using modern notation and terminology. Let p be the unknown probability that a certain event C happens in a single trial. Assume that the only possible values of this parameter are $p_1, \ldots, p_n$. Suppose that C occurs a times in $a + b$ independent trials. Find the conditional probabilities $P(p_i|a)$ of the alternatives p_i given the result a.

Assuming that the prior probabilities $P(H_i)$ of the alternatives are equal, we obtain from (2) and the binomial distribution, after cancellations of a common factor $\binom{a+b}{a}$,

$$P(p_i|a) = \frac{p_i^a(1 - p_i)^b}{\sum_{j=1}^n p_j^a(1 - p_j)^b}. \tag{3}$$

Bayes used the continuous analogue

$$f(p|a) = \frac{p^a(1 - p)^b}{\int_0^1 p^a(1 - p)^b dp}. \tag{4}$$

Hence he regarded p as an rv with a prior distribution that is uniform over the interval $(0,1)$. This clever result is nowadays known as the formula for *inverse probability*. It has evoked innumerable discussions between adherents of different views in probability and statistics. Formula (4) solves Bayes's problem quoted above provided that a uniform prior distribution can be assumed. Integrating from c to d, Bayes found the answer

$$P(c < p < d|a) = \frac{\int_c^d p^a(1-p)^b dp}{\int_0^1 p^a(1-p)^b dp}. \tag{5}$$

The integral in the denominator is equal to $a!b!/(a+b+1)!$; compare 'Symbols and formulas' at the beginning of the book. Thus the answer depends on the incomplete beta function. This function is available in tables nowadays, but for Bayes the integration presented an obstacle which he attacked with great ingenuity and energy; he succeeded in constructing bounds on the integral which may be used for large values of a and b.

(c) *Laplace and Law of Succession*

Pierre Simon de Laplace (1749–1827) is one of the most influential scholars in the history of science; he was the greatest astronomer and the greatest probabilist of his time. His work *Théorie analytique des probabilités* from 1812 is one of the most important books on probability, perhaps the most important single book on this subject published hitherto.

Among many other things, Laplace was much occupied with inverse probability. He tacitly assumed equal prior probabilities in Bayes's formula. This implies that, in modern notation, he used formulas (3) and (4) for inverse probability.

Laplace also considered the average probability P of success in a future trial given a successes in $a+b$ trials. He obtained from (4)

$$P = \frac{\int_0^1 p \cdot p^a(1-p)^b dp}{\int_0^1 p^a(1-p)^b dp} = \frac{a+1}{a+b+2}. \tag{6}$$

Here we have performed two integrations, using the beta function, followed by a reduction.

Laplace used (6) for establishing his much disputed *Law of Succession*, which he applied to the rising of the sun. His argument is as follows (using his own figures). Regard the rising of the sun as a random event which occurs independently each morning with the unknown probability p. Hitherto, the sun has arisen for 5,000 years, or 1,826,213 days. Assuming that p has a uniform prior density over the interval 0 to 1, he takes $a = 1{,}826{,}213$ and $b = 0$ in (6), which leads to the probability

$$P = \frac{1{,}826{,}214}{1{,}826{,}215}$$

that the sun will rise also tomorrow.

'The rising of the sun' is a question discussed not only by Laplace but also by, for example, the philosopher D. Hume and by R. Price in an Appendix to Bayes (1764); see Dale (1991, p. 44). Laplace's contribution can be seen as a part of the perennial discussion of the probabilistic foundations of induction.

5.8   Ménage I

A man and his wife invite $n - 1$ married couples to a dinner party. The n couples are placed at random at a round table with the sole restriction that men and women take alternate seats. What is the probability P_n that none of the men has his wife next to himself, neither to the left nor to the right?

This is the *ménage problem*, formulated at the end of the 19th century. It has been considered a tricky combinatorial problem and an explicit solution was not found until the 1930s. Note that if only those couples are counted where the spouse sits on the right-hand side of her husband, we have the classical rencontre problem discussed in Section 5.4. For the history and other aspects of the ménage and the rencontre problems, see Bogart and Doyle (1986), Holst (1991) and Takács (1981).

Number the seats around the table $1, 2, \ldots, 2n$. Let H_k be the event 'a married couple occupies seats k and $k + 1$'. (If $k = 2n$, the next seat is of course seat 1.)

We note in passing that these events are not exchangeable. For example, $H_1 H_2$ cannot occur, but $H_1 H_3$ can, so $P(H_1 H_2) = 0$ but $P(H_1 H_3) > 0$.

We have

$$P(H_{i_1} H_{i_2} \cdots H_{i_j}) = \frac{(n - j)!}{n!}, \tag{1}$$

whenever all the indices $1 \leq i_1 < \cdots < i_j \leq 2n$ are at least two units apart and $i_j - i_1 \leq 2n - 2$, otherwise this probability is 0.

Example

Let us prove (1) in a special case; the general proof is similar.

Consider a dinner for $n = 4$ couples. Let us determine the probability $P(H_1 H_3)$.

We begin by placing the ladies on odd-numbered chairs 1, 3, 5 and 7. Denote the woman on chair 1 by A and her husband by a, the woman on chair 3 by B and her husband by b, etc. The event $H_1 H_3$ occurs if a is placed on chair 2 and b on chair 4, while c and d are placed on chairs 6 and 8 *in any order*; compare the table.

chair		1	2	3	4	5	6	7	8
women			A		B		C		D
husbands: case 1			a		b		c		d
husbands: case 2			a		b		d		c

Thus there are 2! favourable cases. The number of possible ways to seat the men is 4!, and so we obtain $P(H_1 H_3) = 2!/4! = 1/12$.

The determination of $P(H_i H_j)$ for all odd-numbered H's is similar, and consequently we have proved (1) for $j = 2$.

We shall now show that the probability P_n is given by

$$P_n = \sum_{j=0}^{n} (-1)^j \frac{2n}{2n-j} \binom{2n-j}{j} \frac{(n-j)!}{n!} . \tag{2}$$

Using the inclusion–exclusion formula in Section 2.4, the probability can be written

$$P_n = 1 - S_1 + S_2 - \cdots + S_{2n} , \tag{3}$$

where

$$S_j = \sum_{1 \leq i_1 < \cdots < i_j \leq 2n} P(H_{i_1} \cdots H_{i_j}) . \tag{4}$$

In this sum the only nonzero terms are those where all indices are at least two units apart, and all these are $(n-j)!/n!$ according to (1); note that this implies that $S_j = 0$ for $j > n$. To calculate S_j we need to know the number of nonzero terms in (4). That number is the same as the number of ways of placing j dominos in a ring with $2n$ places, in such a way that each domino occupies two adjacent places and the dominos do not overlap. This can be done in

$$\frac{2n}{2n-j} \binom{2n-j}{j}$$

different ways; for a proof see Holst (1991). Hence it follows that

$$S_j = \frac{2n}{2n-j} \binom{2n-j}{j} \frac{(n-j)!}{n!} , \tag{5}$$

for $j = 0, 1, \ldots, n$. Formulas (3) and (5) imply (2).

Formula (2) with $n = 4$ gives

$$P_4 = 1 - \frac{8}{7}\binom{7}{1}\frac{1}{4} + \frac{8}{6}\binom{6}{2}\frac{1}{12} - \frac{8}{5}\binom{5}{3}\frac{1}{24} + \frac{8}{4}\binom{4}{4}\frac{1}{24} = \frac{1}{12} .$$

In this simple case, P_4 can easily be found directly, using the following table:

chair	1	2	3	4	5	6	7	8
women		A		B		C		D
husbands: case 1		c		d		a		b
husbands: case 2		d		a		b		c

We leave the details to the reader.

We shall discuss another ménage problem in Sections 7.7 and 8.6.

Here is a problem for the reader. Four couples, that is, eight persons in all, are seated at random along one side of a table; men and women take alternate seats. The probability that no couple sits next to each others is 1/8. Show this.

6
Random permutations

As we all know, n elements can be ordered in $n!$ ways. By selecting one of these orderings at random we obtain a *random permutation*. Such permutations are studied intensely nowadays using tools from combinatorics, probability and mathematics; there are also classical problems in this area. The present chapter gives some examples of what has been achieved, and shows the relationship to some other parts of mathematics. For example, we study the number of cycles in a random permutation and demonstrate their close relationships to Stirling numbers of the first kind. The last two sections, which are somewhat more advanced, show that sequences of continuous rv's give rise to problems concerning random permutations as well.

We do not know of any book dealing exclusively with random permutations; the results are scattered over many books and journals. Mathematical books with combinatorial content are also of interest.

6.1 Runs I

A *run* of length r is a sequence of r symbols of one kind preceded and followed by (if anything) symbols of some other kind. For example, in the sequence 1 1 0 0 0 1, there are three runs of lengths 2, 3 and 1.

In this section, we deal with combinatorial runs obtained by permuting a 1's and b 0's randomly. (In Section 14.1 we shall discuss another type of run, obtained, for example, by tossing a coin.) Let X be the number of runs of 1's and Y that of 0's. We are interested in the joint distribution of X and Y, and in the distribution of the total number, $Z = X + Y$.

To find these distributions, we use the classical definition of probability, dividing the number of favourable cases by that of possible cases. At first sight it may seem difficult to enumerate the favourable cases, and it comes as a surprise that it is quite simple.

Let f_{ij} be the number of permutations with i 1-runs and j 0-runs, and f_k the number of permutations with k runs, without regard to type. We have

$$P(X = i, Y = j) = \frac{f_{ij}}{c}; \quad P(Z = k) = \frac{f_k}{c},$$

where
$$c = \binom{a+b}{a}.$$

Example. $a = 2, b = 3$

The sequence 1 1 0 0 0 can be permuted in $c = \binom{5}{2} = 10$ ways:

$$
\begin{array}{ccccc}
1 & 1 & 0 & 0 & 0 \\
1 & 0 & 1 & 0 & 0 \\
1 & 0 & 0 & 1 & 0 \\
1 & 0 & 0 & 0 & 1 \\
0 & 1 & 1 & 0 & 0 \\
0 & 1 & 0 & 1 & 0 \\
0 & 1 & 0 & 0 & 1 \\
0 & 0 & 1 & 1 & 0 \\
0 & 0 & 1 & 0 & 1 \\
0 & 0 & 0 & 1 & 1 \\
\end{array}
$$

The f_{ij}'s are shown in the following table:

$i\backslash j$	1	2	3
1	2	2	0
2	1	4	1

Note that the i, j's for positive entries always satisfy $|i - j| \leq 1$. This is true quite generally (why?). We also show the f_k's:

k	2	3	4	5
f_k	2	3	4	1

Dividing the entries in the tables by 10 we obtain the probability functions of (X, Y) and Z.

We now consider the general case.

(a) *Distribution of* (X, Y)

The a 1's can be partitioned into i groups in

$$\binom{a-1}{i-1}$$

ways. (Divide them into groups by $i - 1$ vertical signs.) Similarly, the b 0's can be partitioned into j groups in

$$\binom{b-1}{j-1}$$

ways. Combining the 1's and the 0's into one sequence, we realize that for $|i - j| = 1$, f_{ij} is the product of these two binomial coefficients, and for $i = j$, it is twice that product; all other f_{ij}'s are zero. Dividing by c we obtain the final result

$$P(X = i, Y = j) = \binom{a-1}{i-1}\binom{b-1}{j-1} \bigg/ \binom{a+b}{a},$$

for all $i = 1, \ldots, a$ and $j = 1, \ldots, b$ such that $|i - j| = 1$, and for $i = j$

$$P(X = i, Y = i) = 2\binom{a-1}{i-1}\binom{b-1}{i-1} \bigg/ \binom{a+b}{a}.$$

(b) *Distribution of Z*

The distribution of the total number, $Z = X + Y$, of runs can be obtained from the distribution in (a) by summation. In particular, when a and b are equal it is found that

$$P(Z = 2k) = 2\binom{a-1}{k-1}\binom{a-1}{k-1} \bigg/ \binom{2a}{a},$$

$$P(Z = 2k+1) = 2\binom{a-1}{k}\binom{a-1}{k-1} \bigg/ \binom{2a}{a}.$$

Here Z can attain the values $2, 3, \ldots, 2a$.

In Section 9.3 we will consider a problem related to combinatorial runs and will then indicate how expectations of X and Y can be found in a simple way, using zero–one rv's.

Finally, we append a problem for the reader. Show that the probability function of the number X of 1-runs is given by

$$P(X = i) = \binom{a-1}{i-1}\binom{b+1}{i} \bigg/ \binom{a+b}{a}$$

for $i = 1, 2, \ldots, \min(a, b+1)$.

The reader is invited to study the *Mississippi problem:* Consider a random permutation of the letters in the word

$$MISSISSIPPI$$

Single version. Show that the probability that no S's stand next to each other is $7/33$.

Double version. Show that the probability that no S's stand next to each other and no I's stand next to each other is $26/385$.

Triple version. Show that the probability that no S's stand next to each other, no I's stand next to each other and no P's stand next to each other is $16/275$.

Of these versions, the first is by far the simplest; the second and the third are hard nuts. Hint for the double and the triple versions: First, order the letters $IIIISSSS$ at random and consider the total number of runs. Second, insert the letters MPP at random. (Teachers who like the problem are advised to replace Mississippi by some word leading to simpler calculations.)

References: David and Barton (1962), Mood (1940).

6.2   Cycles in permutations

(a) *Introduction*

A permutation of $1\,2\,3\,4\,5$, say $2\,1\,3\,5\,4$, can be written as a one-to-one transformation

$$1\,2\,3\,4\,5 \to 2\,1\,3\,5\,4. \tag{1}$$

Alternatively, it can be described by its cycles

$$(1 \to 2 \to 1); \quad (3 \to 3); \quad (4 \to 5 \to 4).$$

In (1) the number 1 is replaced by 2 and 2 by 1, which gives the first cycle, and so on. Hence there are three cycles in this permutation.

Consider a random permutation of $1, 2, \ldots, n$, that is, one of the $n!$ possible permutations of these numbers chosen at random. We denote by X_n the number of cycles in the permutation. We want to find the probability function of X_n:

$$p_n(k) = P(X_n = k)$$

for $k = 1, 2, \ldots, n$.

Example

When $n = 3$, there are $3! = 6$ permutations, two of which have one cycle, three of which have two cycles and one of which has three cycles. Consequently, X_3 assumes the values 1, 2 and 3 with probabilities $p_3(1) = 2/6$, $p_3(2) = 3/6$ and $p_3(3) = 1/6$, respectively.

(b) *A recurrence relation*

Set $p_n(k) = 0$ for $k = 0$ and $k > n$, and also $p_1(1) = 1$. We then have the recurrence relation

$$p_n(k) = \frac{1}{n}p_{n-1}(k-1) + \left(1 - \frac{1}{n}\right)p_{n-1}(k), \qquad (2)$$

where $k = 1, 2, \ldots, n$ and $n = 2, 3, \ldots$. The relation is proven noting that a random permutation of n elements has k cycles if one of the following two disjoint events occurs:

1. The elements $1, 2, \ldots, n-1$ form $k-1$ cycles, and the nth element forms a cycle of its own. The probability of this event is $(1/n)p_{n-1}(k-1)$, since the probability that in a random permutation of n elements the element n forms a cycle of its own is $1/n$.

2. The elements $1, 2, \ldots, n-1$ form k cycles, and the nth element is situated in one of these cycles. The probability of this event is $(1 - 1/n)p_{n-1}(k)$.

Adding the probabilities for the two cases we obtain (2). We write

$$p_n(k) = \frac{1}{n!}\begin{bmatrix} n \\ k \end{bmatrix}, \qquad (3)$$

where $\begin{bmatrix} n \\ k \end{bmatrix}$ denotes the number of permutations with k cycles. The numbers $\begin{bmatrix} n \\ k \end{bmatrix}$ are called *Stirling numbers of the first kind*; see also the next section.

(c) *Representation of X_n as a sum*

In view of (2) we can represent X_n as a sum $X_n = X_{n-1} + U_n$ of the two independent rv's X_{n-1} and U_n, where U_n is 1 with probability $1/n$ and 0 with probability $1 - 1/n$; that is, $U_n \sim \text{Bin}(1, 1/n)$. This implies in its turn that X_n can be written as a sum

$$X_n = U_1 + U_2 + \cdots + U_n \qquad (4)$$

of the independent rv's $U_i, i = 1, 2, \ldots, n$, where $U_i \sim \text{Bin}(1, 1/i)$. Hence, X_n has a Poisson–binomial distribution; see Sections 2.6 and 8.4.

Since $E(U_i) = 1/i$ it follows from (4) that

$$E(X_n) = 1 + \frac{1}{2} + \cdots + \frac{1}{n}.$$

Thus in a long random permutation the mean number of cycles is approximately $\ln n + \gamma$, where γ is Euler's constant (see 'Symbols and formulas' at the beginning of the book). We leave it to the reader to show that for large n we have $Var(X_n) \approx \ln n + \gamma - \pi^2/6$.

6.3   Stirling numbers of the first kind

Stirling numbers arise in a wide variety of applications. In Section 5.6 we considered the Stirling number of the second kind $\{^n_k\}$. In this section we will study Stirling numbers of the first kind. For a fuller discussion on such numbers we refer to Graham, Knuth and Patashnik (1989), as well as Knuth (1992).

(a) First definition

We already denoted the number of permutations of the set $\{1, 2, \ldots, n\}$ with k cycles by $[^n_k]$ in the preceding section and called it a *Stirling number of the first kind*. This is our first definition.

The notation $[^n_k]$ is not standard; see the interesting paper by Knuth (1992) concerning the history of the notations and related topics.

(b) Second definition

Inserting (3) of the preceding section into (2) of the same section, we get the recursive relation

$$\begin{bmatrix} n \\ k \end{bmatrix} = \begin{bmatrix} n-1 \\ k-1 \end{bmatrix} + (n-1)\begin{bmatrix} n-1 \\ k \end{bmatrix}, \tag{1}$$

which holds for $k = 1, 2, \ldots, n$ and $n = 2, 3, \ldots$, provided that we set $[^1_1] = 1$ and $[^n_k] = 0$ for $k = 0$ or $k > n$. This formula is taken as our second definition of Stirling numbers of the first kind. Using this recursion the numbers are easily computed; see Table 1.

Table 1. Stirling numbers of the first kind $[^n_k]$.

$n \backslash k$	1	2	3	4	5	6	7	8
1	1							
2	1	1						
3	2	3	1					
4	6	11	6	1				
5	24	50	35	10	1			
6	120	274	225	85	15	1		
7	720	1,764	1,624	735	175	21	1	
8	5,040	13,068	13,132	6,769	1,960	322	28	1

(c) Third definition

We saw in the preceding section that we can represent X_n as a sum of independent rv's:

$$X_n = U_1 + \cdots + U_n,$$

where $U_i \sim \text{Bin}(1, 1/i)$. Since the probability generating function of U_i is

$$E(s^{U_i}) = s \cdot \frac{1}{i} + 1 \cdot \left(1 - \frac{1}{i}\right) = \frac{s + i - 1}{i},$$

that of X_n becomes

$$\frac{1}{n!}\left[s(s+1)\cdots(s+n-1)\right].$$

On the other hand, using (3) of the preceding section, the probability generating function can be written

$$E(s^{X_n}) = \sum_{k=1}^{n} s^k p_n(k) = \frac{1}{n!} \sum_{k=1}^{n} \begin{bmatrix} n \\ k \end{bmatrix} s^k.$$

Hence we obtain

$$s(s+1)\cdots(s+n-1) = \sum_{k=1}^{n} \begin{bmatrix} n \\ k \end{bmatrix} s^k. \tag{2}$$

This relation, which expresses *ascending* or *rising* factorials in terms of *ordinary* powers, can be seen as a third definition of the numbers.

Replacing s with $-s$ in (2) we obtain the relation

$$s(s-1)\cdots(s-n+1) = \sum_{k=1}^{n}(-1)^{n-k} \begin{bmatrix} n \\ k \end{bmatrix} s^k, \tag{3}$$

expressing *descending* or *falling* factorials in ordinary powers. The signed numbers $(-1)^{n-k}\begin{bmatrix} n \\ k \end{bmatrix}$ are sometimes called Stirling numbers of the first kind instead of $\begin{bmatrix} n \\ k \end{bmatrix}$. Stirling's motivation for studying these numbers was the identity (3); the combinatorial and probabilistic interpretations were developed much later (see Knuth (1992)).

(d) *Another application*

Let X be an rv with the ascending factorial moments

$$E[X(X+1)\cdots(X+n-1)]$$

and the descending factorial moments

$$E[X(X-1)\cdots(X-n+1)].$$

It follows from (2) and (3) that these moments can be expressed in terms of ordinary moments in the following way:

$$E[X(X+1)\cdots(X+n-1)] = \sum_{k=1}^{n} \begin{bmatrix} n \\ k \end{bmatrix} E(X^k), \tag{4}$$

$$E[X(X-1)\cdots(X-n+1)] = \sum_{k=1}^{n} (-1)^{n-k} \begin{bmatrix} n \\ k \end{bmatrix} E(X^k). \tag{5}$$

Recall that in Section 5.6, Subsection (c), the ordinary moments were expressed in factorial moments using Stirling numbers of the second kind.

We leave it to the reader to show that the probability P_n that all persons in a group of n have different birthdays can be expressed as

$$P_n = \sum_{k=1}^{n} \begin{bmatrix} n \\ k \end{bmatrix} (-d)^{k-n}.$$

Here d is the number of days; see also Section 7.1.

6.4   Ascents in permutations

(a) *Introduction*

Consider a permutation of the integers $1\,2\,3\,4\,5\,6$, say $2\,1\,3\,5\,4\,6$. As $1 < 3$, $3 < 5$, $4 < 6$ there are three *ascents* in this permutation.

Let the number of ascents in a random permutation of $1, 2, \ldots, n$ be X_n. Clearly, we have $0 \le X_n \le n-1$. Let

$$p_n(k) = P(X_n = k)$$

for $k = 0, 1, \ldots, n-1$ be the probability function of the rv X_n. We shall determine this function.

(b) *A recursion formula*

We shall derive the recursive relation

$$p_n(k) = \frac{k+1}{n} p_{n-1}(k) + \frac{n-k}{n} p_{n-1}(k-1), \tag{1}$$

which holds for $k = 1, 2, \ldots, n-1$ and $n = 2, 3, \ldots$, provided we set $p_n(0) = 1/n!$ and $p_n(k) = 0$ when $k \ge n$.

The recursion (1) is proved in the following way: Let $\rho = (\rho_1, \ldots, \rho_{n-1})$ be a random permutation of $1, 2, \ldots, n-1$. A random permutation π of

$1, 2, \ldots, n$ is obtained by selecting a position j at random, say, for the element n:

$$\pi = (\rho_1, \ldots, \rho_{j-1}, n, \rho_j, \ldots, \rho_{n-1}).$$

There are four cases:

1. $j = 1$: The number of ascents in π is the same as that in ρ.
2. $2 \leq j \leq n - 1$ and $\rho_{j-1} < \rho_j$: The number of ascents in π is the same as that in ρ.
3. $2 \leq j \leq n - 1$ and $\rho_{j-1} > \rho_j$: The number of ascents in π is one greater than that in ρ.
4. $j = n$: The number of ascents in π is one greater than that in ρ.

Each position of the element n occurs with probability $1/n$. Summing the four alternatives, we obtain

$$p_n(k) = \frac{1}{n}p_{n-1}(k) + \frac{k}{n}p_{n-1}(k) + \frac{n-1-k}{n}p_{n-1}(k-1) + \frac{1}{n}p_{n-1}(k-1),$$

which leads to (1).

Formula (1) can, of course, be used for determining the probability function of X_n for any n. However, it is more convenient to use a table of Eulerian numbers; see the next section.

The famous nineteenth-century American astronomer *Simon Newcomb* amused himself by playing a game of solitaire related to the following question: A deck of cards contains n different cards, say numbered $1, 2, \ldots, n$. The n cards are drawn at random and are put into a pile as long as the card drawn is lower than its predecessor; a new pile is started when a higher card is drawn. We leave it to the reader to show that

a. the number of piles has the same distribution as $1 + X_n$,
b. the expected number of piles is $(n + 1)/2$. (Hint: use zero–one rv's.)

A good reference to Simon Newcomb's problem is Barton and Mallows (1965).

6.5   Eulerian numbers

This is a continuation of the preceding section. As before, let the probability function of X_n be $p_n(k)$ for $k = 0, \ldots, n - 1$. We write

$$p_n(k) = \frac{1}{n!}\left\langle {n \atop k} \right\rangle, \tag{1}$$

where $\left\langle {n \atop k} \right\rangle$ is the number of permutations of $1, 2, \ldots, n$ with k ascents. These numbers are called *Eulerian numbers*. (They should not be confused with the Euler numbers mentioned in Section 6.9.) By symmetry it follows that

$$\left\langle {n \atop k} \right\rangle = \left\langle {n \atop n-k-1} \right\rangle.$$

Inserting (1) in (1) of the preceding section we obtain the recursion formula

$$\left\langle {n \atop k} \right\rangle = (k+1)\left\langle {n-1 \atop k} \right\rangle + (n-k)\left\langle {n-1 \atop k-1} \right\rangle. \tag{2}$$

This formula holds for $k = 1, 2, \ldots, n-1$ and $n = 2, 3, \ldots$, provided we set $\left\langle {n \atop 0} \right\rangle = 1$ and $\left\langle {n \atop k} \right\rangle = 0$ when $k \geq n$.

Table 1. Eulerian numbers $\left\langle {n \atop k} \right\rangle$.

$n \backslash k$	0	1	2	3	4	5	6	7
1	1							
2	1	1						
3	1	4	1					
4	1	11	11	1				
5	1	26	66	26	1			
6	1	57	302	302	57	1		
7	1	120	1,191	2,416	1,191	120	1	
8	1	247	4,293	15,619	15,619	4,293	247	1

It also deserves to be mentioned that Eulerian numbers are explicitly given by

$$\left\langle {n \atop k} \right\rangle = \sum_{i=0}^{k} (-1)^i \binom{n+1}{i}(k+1-i)^n.$$

In particular we have

$$\left\langle {n \atop 1} \right\rangle = 2^n - n - 1.$$

See Graham, Knuth and Patashnik (1989, p. 255).

6.6 Exceedances in permutations

Let $(\pi_1, \pi_2, \ldots, \pi_n)$ be a random permutation of $1, 2, \ldots, n$. In the two preceding sections we saw that the number X_n of ascents in this permutation has the probability function

$$p_n(k) = \frac{1}{n!}\left\langle {n \atop k} \right\rangle$$

and we proved the recursion

$$p_n(k) = \frac{k+1}{n} p_{n-1}(k) + \frac{n-k}{n} p_{n-1}(k-1). \tag{1}$$

We say that j is an *exceedance* if $\pi_j > j$. Let Z_n be the number of exceedances in a random permutation of $1, 2, \ldots, n$. Consider a random permutation $(\rho_1, \rho_2, \ldots, \rho_{n-1})$ of $1, 2, \ldots, n-1$ with Z_{n-1} exceedances. If in the permutation $(\rho_1, \rho_2, \ldots, \rho_{n-1}, n)$ the n is exchanged with an element at a random place, including itself, we obtain a random permutation of $1, 2, \ldots, n$. From this we can easily see that

$$P(Z_n = k) = \frac{1}{n} P(Z_{n-1} = k) + \frac{k}{n} P(Z_{n-1} = k)$$
$$+ \frac{n-1-(k-1)}{n} P(Z_{n-1} = k-1),$$

or, equivalently,

$$P(Z_n = k) = \frac{k+1}{n} P(Z_{n-1} = k) + \frac{n-k}{n} P(Z_{n-1} = k-1), \tag{2}$$

with the starting value $P(Z_1 = 1) = 1$. As (1) and (2) are identical recursions it follows that rv's X_n and Z_n have the same distributions. This shows that

$$P(Z_n = k) = \frac{1}{n!} \left\langle {n \atop k} \right\rangle,$$

where $\left\langle {n \atop k} \right\rangle$ are the Eulerian numbers introduced in the preceding section. This is a remarkable result showing that the number of permutations having k ascents is the same as the number having k exceedances.

Let $G_n(s)$ be the probability generating function of Z_n. By an argument similar to the one above one can show that

$$G_n(s) = \frac{s(n-1)+1}{n} G_{n-1}(s) + \frac{s(1-s)}{n} G'_{n-1}(s).$$

We leave this rather difficult problem to the reader.

6.7   Price fluctuations

An economist employed at a chocolate factory studies the price fluctuations of cacao from day to day. He is interested in the lengths of sequences of increasing prices. Since the resulting time series is difficult to analyze because of the nonstationarity of prices and the dependence at short time distances, he resolves to neglect these complications as a first approximation.

Consider to this end iid continuous rv's $X_1, X_2, \ldots$. If $X_1 < X_2$, the sequence begins with an increase. Let in this case N be the number of increases until the first decrease, that is, N is the smallest integer k such that

$$X_1 < X_2 < \cdots < X_{k+1} > X_{k+2}.$$

On the other hand, if $X_1 > X_2$, we take $N = 0$, which, of course, happens with probability $\frac{1}{2}$. By symmetry it follows that the $n!$ possible orderings of $X_1, X_2, \ldots, X_n$ occur with the same probability $1/n!$. For example, if $n = 3$, there are $3! = 6$ possible orderings:

$$X_1 < X_2 < X_3; \qquad X_1 < X_3 < X_2;$$
$$X_2 < X_1 < X_3; \qquad X_2 < X_3 < X_1;$$
$$X_3 < X_1 < X_2; \qquad X_3 < X_2 < X_1.$$

We shall now determine the probability for k consecutive increases, that is, the probability

$$P(N = k) = P(X_1 < X_2 < \cdots < X_{k+1} > X_{k+2})$$

for $k = 1, 2, \ldots$. We can see that, for $k \geq 1$,

$$P(N = k) = P(X_1 < \cdots < X_{k+1}) - P(X_1 < \cdots < X_{k+2})$$
$$= \frac{1}{(k+1)!} - \frac{1}{(k+2)!} = \frac{k+1}{(k+2)!},$$

and this result holds for $k = 0$ as well. This is the probability function of N.

The interested reader may perhaps show that the mean is $e - 2$.

6.8 Oscillations I

The following two sections are meant for mathematically minded readers.

Let $c_1, c_2, \ldots$ be given numbers that oscillate. We distinguish between two types of oscillations:

Type 1: $c_1 > c_2 < c_3 > \cdots$; \qquad Type 2: $c_1 < c_2 > c_3 < \cdots$.

Let $X_1, X_2, \ldots$ be a sequence of iid continuous rv's. Collect these variables as long as the sequence oscillates as Type 1:

$$X_1 > X_2 < X_3 > X_4 < \cdots .$$

The stopping time N is the subscript of the first rv that violates this rule. For example, if $X_1 > X_2 < X_3 < X_4$, we have $N = 4$. We want to find the probability function

$$p_N(k) = P(N = k), \qquad k = 2, 3, \ldots,$$

of N and the mean $E(N)$.

It is convenient first to determine the cumulative probabilities

$$P_N(k) = P(N > k)$$

for $k = 0, 1, \ldots$. We have, of course, $P_N(0) = P_N(1) = 1$. Clearly, the distribution of N does not depend on the distribution of the rv's, but only on their order.

As an example, consider the cases $N > 2$ and $N > 3$. We have $N > 2$ if $X_1 > X_2$, so $P_N(2) = \frac{1}{2}$. Furthermore, we have $N > 3$ if $X_1 > X_2 < X_3$. Two orderings are favourable, namely $X_2 < X_1 < X_3$ and $X_2 < X_3 < X_1$; hence $P_N(3) = 2/3! = 1/3$.

We shall now determine $P_N(k)$ for a general k. The rv's $X_1, \ldots, X_k$ can be ordered in $k!$ ways. Let a_k be the number of orderings such that the rv's oscillate as

$$X_1 > X_2 < X_3 > \cdots > X_k \tag{1}$$

if k is even and

$$X_1 > X_2 < X_3 > \cdots < X_k \tag{2}$$

if k is odd. It follows that

$$P_N(k) = \frac{a_k}{k!}. \tag{3}$$

The a's can be found as follows:

Let n_{jk} be the number of orderings satisfying (1) or (2) such that the last rv X_k has rank j counted from the bottom. Then we have

$$a_k = \sum_{j=1}^{k} n_{jk}.$$

We can see that $n_{11} = n_{12} = 1$, that $n_{kk} = 0$ if k is even and at least 2 and that $n_{1k} = 0$ if k is odd and at least 3.

Using these starting values, the numbers n_{jk} can be found recursively in the following way:

$$n_{jk} = \sum_{i > j} n_{i, k-1}$$

if k is even and at least 2, and

$$n_{jk} = \sum_{i<j} n_{i,k-1}$$

if k is odd and at least 3. This can be explained as follows: Suppose that k is even so that (1) holds and suppose that X_k has rank j counted from the bottom. Each oscillation in (1) has $X_{k-1} > X_k$. Hence to each such oscillation there corresponds an oscillation of $X_1, \ldots, X_{k-1}$, where X_{k-1} has rank i with $i > j$. When k is odd, we have $i < j$ instead.

It is convenient to find the numbers n_{jk} using Table 1. These numbers are shown within the table. Each row is obtained from the preceding one by cumulative summation, from left to right in odd-numbered rows and from right to left in even-numbered rows. The numbers a_k are obtained by summation in each row.

Table 1. Generation of the numbers n_{jk} and a_k (see text).

k								a_k
1							1	1
2						1	0	1
3					0	1	1	2
4				2	2	1	0	5
5			0	2	4	5	5	16
6		16	16	14	10	5	0	61
7	0	16	32	46	56	61	61	272

It follows from (3) and Table 1 that

$$P_N(1) = \frac{1}{1!}; \quad P_N(2) = \frac{1}{2!}; \quad P_N(3) = \frac{2}{3!}; \quad P_N(4) = \frac{5}{4!}$$

and so on. Since $p_N(k) = P_N(k-1) - P_N(k)$, we find successively

$$p_N(2) = P_N(1) - P_N(2) = 1 - \tfrac{1}{2} = \tfrac{1}{2},$$
$$p_N(3) = P_N(2) - P_N(3) = \tfrac{1}{2} - \tfrac{1}{3} = \tfrac{1}{6},$$
$$p_N(4) = P_N(3) - P_N(4) = \tfrac{1}{3} - \tfrac{5}{24} = \tfrac{1}{8}$$

and so on.

The mean of the distribution can be found from either of the expressions

$$E(N) = \sum_{k=0}^{\infty} k\, p_N(k)$$

or

$$E(N) = \sum_{k=0}^{\infty} P_N(k).$$

Extending Table 1 to higher values than $k = 7$, we find that, to two decimal places, $E(N) \approx 3.41$.

6.9   Oscillations II

In the preceding section we introduced two types of oscillations for a sequence of numbers. We shall derive an exact value of the mean length $E(N)$ of Type 1 oscillations

$$X_1 > X_2 < X_3 > \cdots$$

of iid continuous rv's $X_1, X_2, \ldots$, where N is the first index that violates the rule of a Type 1 oscillation. Let $P_N(k) = P(N > k)$ as before. In the preceding section we obtained these probabilities from Table 1 in the way described in that section. Now we shall derive a recursive relation which directly determines these probabilities:

$$P_N(k+1) = \frac{1}{2(k+1)} \sum_{j=0}^{k} P_N(j) P_N(k-j) \tag{1}$$

for $k = 1, 2, \ldots$, starting from $P_N(0) = P_N(1) = 1$.

Consider the probability $2P_N(k+1)$ that the sequence $X_1, \ldots, X_{k+1}$ oscillates as Type 1 or Type 2. (Clearly, Type 1 and Type 2 occur with the same probability.) Suppose that X_{j+1} is the smallest of $X_1, \ldots, X_{k+1}$, where j can attain one of the values $0, 1, \ldots, k$.

Let us first illustrate what happens in two special cases:

1. $k + 1 = 6$, $j = 3$: The oscillations of $X_1, \ldots, X_6$ is of Type 1:

$$X_1 > X_2 < X_3 > X_4 < X_5 > X_6.$$

Both X_1, X_2, X_3 and X_5, X_6 have oscillations of Type 1 in this case.

2. $k + 1 = 6$, $j = 2$: The oscillations of $X_1, \ldots, X_6$ are of Type 2:

$$X_1 < X_2 > X_3 < X_4 > X_5 < X_6.$$

Now X_1, X_2 oscillate as Type 2 and X_4, X_5, X_6 as Type 1.

Generally, $X_1, X_2, \ldots, X_{k+1}$ oscillate as Type 1 or Type 2 if one of the following events H_{j+1}, $j = 0, 1, \ldots, k$, occurs. We define H_{j+1} as the joint event described in (a), (b) and (c):

a. X_{j+1} is the smallest rv,
b. $X_1, \ldots, X_j$ oscillate (depending on j) as Type 1 or Type 2,
c. $X_{j+2}, \ldots, X_{k+1}$ oscillate (depending on j) as Type 1 or Type 2.

By summing over the disjoint events H_{j+1} we obtain the relation

$$2P_N(k+1) = \frac{1}{k+1} \sum_{j=0}^{k} P_N(j)P_N(k-j).$$

This is equivalent to (1).

We now introduce the generating function

$$G(t) = \sum_{i=0}^{\infty} P_N(i)t^i \tag{2}$$

of the probabilities $P_N(i)$. We shall use the relations (1) for determining an explicit expression for this function. Squaring $G(t)$ and using (1) we obtain

$$[G(t)]^2 = \left[\sum_{i=0}^{\infty} P_N(i)t^i \right]^2 = \sum_{k=0}^{\infty} \left[\sum_{j=0}^{k} P_N(j)P_N(k-j) \right] t^k$$

$$= 1 + 2\sum_{k=1}^{\infty}(k+1)P_N(k+1)t^k = 1 + 2\sum_{i=2}^{\infty} iP_N(i)t^{i-1}.$$

Observing that

$$G'(t) = \sum_{i=1}^{\infty} iP_N(i)t^{i-1}$$

and $P_N(1) = 1$, we obtain the relation

$$2G'(t) = 1 + [G(t)]^2.$$

This differential equation can be rewritten as

$$\frac{dG(t)}{1 + [G(t)]^2} = \frac{dt}{2}.$$

The general solution is

$$\arctan G(t) = t/2 + C$$

or

$$G(t) = \tan(t/2 + C).$$

Now $G(0) = P_N(0) = 1$. Hence,

$$G(t) = \tan\left(\frac{t}{2} + \frac{\pi}{4}\right). \tag{3}$$

We may use this function for determining $E(N)$. As seen from (2), all we need to do is to set $t = 1$ in (3), and the result is

$$E(N) = \tan\left(\frac{1}{2} + \frac{\pi}{4}\right) \approx 3.4082.$$

It also follows that

$$Var(N) = 1 + E(N).$$

Let us add a mathematical remark. It follows from (3) that

$$G(t) = \frac{1 + \tan(t/2)}{1 - \tan(t/2)} = \frac{1}{\cos t} + \tan t.$$

Hence when $G(t)$ is expanded in a power series, we obtain the sum of the familiar expansions

$$\frac{1}{\cos t} = \sum_{i=0}^{\infty} \frac{E_{2i}}{(2i)!} t^{2i} ; \quad \tan t = \sum_{i=0}^{\infty} \frac{T_{2i+1}}{(2i+1)!} t^{2i+1}.$$

Here $E_0 = 1$, $E_2 = 1$, $E_4 = 5$, $E_6 = 61$, etc. are the *Euler numbers* and $T_1 = 1$, $T_3 = 2$, $T_5 = 16$, $T_7 = 272$, etc. are the *tangent numbers*. A look at (2) will tell the reader that the numbers a_i in

$$P_N(i) = \frac{a_i}{i!}$$

for $i = 0, 1, \ldots$ are obtained by combining the E's and the T's. In fact, we find (compare Table 1 in the preceding section) that

$$a_1 = T_1; \quad a_2 = E_2; \quad a_3 = T_3; \quad a_4 = E_4$$

and so on. It is clear that the Euler numbers and the tangent numbers are integers. (Why?)

7

Miscellaneous I

In this chapter and in Chapter 9, we present an assortment of problems and snapshots from different ages. We discuss the birthday problem, the rencontre problem and the ménage problem, the latter appearing in a different version than in Section 5.8. Also, we consider coupon collecting, records, poker and blackjack, among other things.

We do not know of any book covering all the subjects discussed in these two chapters, but as usual we recommend Feller (1968) for further reading. Special references are given in some sections.

7.1   Birthdays

When two people happen to have the same birthday we say that a *coincidence* occurs.

Problem 1. Probability of a common birthday

This is an 'old nut' cracked in innumerable textbooks: Select 23 people at random. Find the probability that at least two of them have the same birthday.

Assume that each year has d days, that the birth rate is the same throughout the year and that all people are born independently of each other. This means, for example, that a given person is born on day i with probability $1/d$.

Take generally n people, where $2 \leq n \leq d$. The probability p_n that the people have different birthdays is given by

$$p_n = \frac{d-1}{d} \cdot \frac{d-2}{d} \cdots \frac{d-n+1}{d} = \binom{d}{n} \frac{n!}{d^n}. \tag{1}$$

[The probability that the second person was born on a different day than the first is $(d-1)/d$. The probability that the third person was born on a different day than the first and the second person (given that the second person was born on a different day the the first) is $(d-2)/d$, and so on.] The probability we seek is $1 - p_n$.

Using a pocket calculator we find that, for $d = 365$, we have $1 - p_{23} \approx 0.5073$. For an approximate way to compute this probability, see Section 8.1. It is amazing that the probability is so large. For $n = 30$ the probability is about 0.71 and for $n = 40$ about 0.89. Thus it is rather safe to bet that, among 40 people, at least 2 have the same birthday.

Problem 2. Expected number to get a common birthday

Select one person at a time until for the first time a person is obtained with the same birthday as a previously selected one. Let N be the number of selected people. Find $E(N)$.

The event 'the first n selected people have different birthdays' is the same as the event '$N > n$'. Therefore $P(N > n)$ is given by (1). Using (2) in Section 3.1 we get

$$E(N) = \sum_{n=0}^{\infty} P(N > n) = \sum_{n=0}^{d} \binom{d}{n} \frac{n!}{d^n}. \tag{2}$$

A numerical calculation for $d = 365$ gives $E(N) \approx 24.617$. We have, approximately,

$$E(N) \approx \sqrt{\frac{\pi}{2} \cdot 365} + \frac{2}{3} \approx 24.611.$$

This formula will be proved in Section 15.2 and applied to the birthday problem in Section 15.3.

Show that, in a group of 23 Martians, the probability that at least 2 have the same birthday is 0.3109. (A Martian year has 687 days.)

7.2   Poker

As already mentioned in Section 4.1, Cardano 'discovered' the classical rule $p = f/c$ for determining probabilities. Here f is the number of favourable cases and c the number of possible cases. Since Cardano's time there have been countless applications of this formula, many well founded, others less so. In games involving coins, dice, cards and urns, the formula has its right place, and it is used even nowadays by friends of combinatorial probability. In this section, we shall apply it to poker.

As is well known, there are 52 cards in an ordinary deck of cards, consisting of 4 suits with 13 denominations in each. A poker deal contains 5 randomly selected cards. A 'full house' means that the player receives 3 cards of one denomination and 2 cards of another denomination; 'three-of-a-kind' means that he gets 3 cards of one denomination, 1 of another denomination and 1 of a third denomination. We shall determine the probability of getting a full house and the probability of getting three-of-a-kind.

(a) *Full house*

The number of possible poker deals is $\binom{52}{5}$. The favourable cases are found as follows: The denominations for the 3 cards and the 2 cards can be selected in $13 \cdot 12$ ways. There are $\binom{4}{3}$ ways of selecting 3 cards from 4 cards with the same denomination; analogously, there are $\binom{4}{2}$ ways of taking out 2 cards. Hence the number of favourable cases is

$$13 \cdot 12 \cdot \binom{4}{3} \cdot \binom{4}{2},$$

and the probability we want becomes

$$\frac{13 \cdot 12 \cdot \binom{4}{3}\binom{4}{2}}{\binom{52}{5}} = \frac{6}{4,165} \approx 0.0014.$$

Now add a joker to an ordinary deck of cards; this is a 'wild card' that can be interpreted as any other card. If a poker hand is taken from this deck, the probability of a full house is $144/44{,}149 \approx 0.0033$. This means that the chance of a full house is more than doubled in the presence of a joker.

(b) *Three-of-a-kind*

The denominations for the 3 cards and the 1 cards can be chosen in $13\binom{12}{2}$ ways (not $13 \cdot 12 \cdot 11$ ways!). Hence we find, in about the same ways as in (a), the probability of getting three-of-a-kind:

$$\frac{13 \cdot \binom{12}{2}\binom{4}{3}\binom{4}{1}\binom{4}{1}}{\binom{52}{5}} = \frac{88}{4,165} \approx 0.0211.$$

This expression is a multivariate hypergeometric probability; see Section 5.1.

(c) *Poker with a die*

Poker may be played with a die instead of a deck of cards. A deal is then obtained by throwing the die five times. Then the probabilities of getting full house or three-of-a-kind are $25/648$ and $25/162$, respectively.

7.3   Negative binomial

An rv X with probability function

$$P(X = k) = \binom{-N}{k}(-P)^k Q^{-N-k}, \tag{1}$$

where $k = 0, 1, \ldots$ and $P > 0, Q = 1 + P$ (note: positive sign!), is said to have a *negative binomial distribution* with parameters N and P.

The quantity $\binom{-N}{k}$ is defined formally in the same way as an ordinary binomial coefficient. For example, we mean by $\binom{-N}{2}$ the quantity $(-N)(-N-1)/2$, which can be written as $\binom{N+1}{2}$. Similarly, we have $\binom{-N}{3} = (-N)(-N-1)(-N-2)/3!$, which can be rewritten as $-\binom{N+2}{3}$, and so on. Changing the binomial coefficients in this way we obtain

$$P(X = k) = \binom{N + k - 1}{k} P^k Q^{-N-k}. \tag{2}$$

The representation (1) explains the name of the distribution: we have replaced the parameters n and p in the binomial distribution by $-N$ and $-P$, respectively.

The negative binomial distribution has mean NP and variance NPQ. (Note the analogy with the moments of the binomial distribution by changing the mean and variance of this distribution in the way indicated above.)

The negative binomial distribution has several applications, one of which is the following: Perform independent trials, where in each trial an event H occurs with probability p and hence its complement H^* occurs with probability $q = 1 - p$. Consider the number W of H^* which occur *before* the moment when H occurs the rth time. The rv W has the probability function

$$P(W = k) = \binom{r + k - 1}{k} p^r q^k,$$

where $k = 0, 1, \ldots$. Since $p + q = 1$, this can be rewritten in the form

$$P(W = k) = \binom{r + k - 1}{k}\left(\frac{q}{p}\right)^k \left(1 + \frac{q}{p}\right)^{-r-k}.$$

As seen from (2), this is the probability function of a negative binomial distribution with parameters $N = r$ and $P = q/p$.

The negative binomial distribution is usually ascribed to Montmort, who used it in 1713 when solving the division problem; see Section 4.3. Hald (1990, p. 61) points out that the distribution appeared already in 1654 in Fermat's analysis of the division problem in his famous correspondence with Pascal ; however, Fermat only considered the special case $p = \frac{1}{2}$.

7.4 Negative hypergeometric I

An urn contains a white and b black balls. Balls are drawn at random without replacement, one at a time.

(a) *Waiting for the first white ball*

Let us draw balls until the first white ball is obtained. The number N of drawings is an rv with probability function

$$P(N = k) = \binom{a+b-k}{a-1} \Big/ \binom{a+b}{a}, \tag{1}$$

for $k = 1, 2, \ldots, b+1$. In fact, we have by the rules of probability

$$P(N = k) = \frac{b}{a+b} \cdot \frac{b-1}{a+b-1} \cdots \frac{b-k+2}{a+b-k+2} \cdot \frac{a}{a+b-k+1}.$$

Rewriting this expression we obtain (1). Alternatively, (1) follows from the general proof in (b) below.

The expectation of N is given by

$$E(N) = \frac{a+b+1}{a+1}. \tag{2}$$

We may use (1) for a direct proof of (2), but shall instead sketch a method which gives insight into the 'mechanism' when drawing without replacement.

Number the black balls from 1 to b. Write

$$N = U_1 + U_2 + \cdots + U_b + 1, \tag{3}$$

where $U_i = 1$ if the ith black ball is drawn before the first white ball, and $U_i = 0$ otherwise.

Now imagine that all balls are drawn from the urn. We then obtain a sequence of $a + b$ balls

$$\ldots W \ldots W \ldots \ldots W \ldots$$

where we have only shown the a white balls. They divide the black balls into $a + 1$ subsequences, each consisting of between 0 and b black balls. It emerges to a reader with intuition (and can be strictly proved using exchangeable events; see Section 2.2) that the ith black ball belongs to any such subsequence with the same probability $1/(a+1)$. Hence

$$P(U_i = 1) = \frac{1}{a+1}.$$

As a consequence all U's have the mean $1/(a+1)$. Using (3) we find

$$E(N) = \frac{b}{a+1} + 1 = \frac{a+b+1}{a+1},$$

and (2) is proved.

Remark

We sometimes exclude the last drawing and only count the number N' of black balls drawn before the first white ball is obtained. We then have, of course, $N' = N - 1$. It is easy to modify the probability function (1), but we only give the mean

$$E(N') = \frac{b}{a+1}. \tag{4}$$

(b) *Waiting for the rth white ball*

Let us generalize (a) by drawing balls until the rth $(1 \leq r \leq a)$ white ball turns up. The required number, N, of balls drawn has probability function

$$P(N = k) = \binom{k-1}{r-1}\binom{a+b-k}{a-r} \Big/ \binom{a+b}{a}, \tag{5}$$

for $k = r, r+1, \ldots, b+r$.

   To prove this, denote a white ball by 1 and a black ball by 0 and consider the $\binom{a+b}{a}$ possible sequences of a 1's and b 0's obtained when drawing all balls from the urn. The favourable sequences are those which have $r - 1$ 1's somewhere in the first $k - 1$ positions, one 1 in the kth position and $a - r$ 1's somewhere in the last $a + b - k$ positions. According to the multiplication principle, their number is seen to be

$$\binom{k-1}{r-1}\binom{1}{1}\binom{a+b-k}{a-r}.$$

This leads to the expression (5).

   The distribution in (5) is called the *negative hypergeometric distribution*. Its mean is given by

$$E(N) = \frac{r(a+b+1)}{a+1}. \tag{6}$$

Remark

It is, of course, possible to count only the black balls drawn before the rth white ball is obtained. We leave it as an exercise to the reader to derive the distribution of the number N' of these balls. Since $N' = N - r$ it follows from (6) that the mean is given by

$$E(N') = \frac{rb}{a+1}.$$

(c) *A special problem*

For later purposes we shall consider a special problem leading to a truncation of the negative hypergeometric distribution.

Consider an urn with r white and r black balls. Draw balls until the rth white or the rth black ball is obtained, whichever comes first. Let N be the number of drawings.

We shall show that N has probability function

$$P(N = k) = 2\binom{k-1}{r-1} \bigg/ \binom{2r}{r} \tag{7}$$

for $k = r, r+1, \ldots, 2r-1$. To obtain this probability function, we observe that there will be k drawings if either (i) $r - 1$ white balls are drawn in the first $k - 1$ drawings and then the last white ball is drawn in the kth drawing, or (ii) a similar event occurs for black balls. Thus there are

$$2\binom{k-1}{r-1}\binom{1}{1}\binom{2r-k}{0} = 2\binom{k-1}{r-1}$$

favourable cases. Since there are $\binom{2r}{r}$ possible cases, we obtain the expression for the probability function given above. We shall later determine the mean of the distribution; see Section 15.8.

7.5   Coupon collecting I

In a food package there is a prize which can be of r different types. All types are distributed at random with the same probability. A person buys one package at a time until he gets a complete set of prizes, that is, at least one of each type. Let N be the number of packages which he has bought. We shall determine the expectation and the probability distribution of N.

Problems of this kind, often referred to as *coupon collector's problems*, will be studied in the present section and in Sections 7.6 and 15.4.

(a) *The expectation of N*

In order to find the expectation of N, we use a trick. Write

$$N = Y_1 + Y_2 + \cdots + Y_r,$$

where Y_j is the number of packages the person must buy in order to increase his collection from $j - 1$ to j different types. We have

$$E(N) = E(Y_1) + E(Y_2) + \cdots + E(Y_r).$$

Suppose that the person has $j-1$ types so that $r-j+1$ are missing. The number of packages which he must buy in order to get one more type has the probability function

$$P(Y_j = k) = (1 - p_j)^{k-1} p_j$$

for $k = 1, 2, \ldots$. Here p_j is the probability of obtaining a new prize. Since the person already has $j-1$ prizes and there are r different types, this probability is equal to $[r - (j-1)]/r$. It follows that

$$E(Y_j) = \frac{1}{p_j} = \frac{r}{(r-j+1)},$$

and hence

$$E(N) = r\left(\frac{1}{r} + \frac{1}{r-1} + \cdots + 1\right).$$

For a large r this is approximately equal to $r(\ln(r) + \gamma)$, where γ is Euler's constant (see 'Symbols and formulas' at the beginning of the book). When $r = 100$ the approximation for the mean gives 518.24. The exact value to two decimal places is 518.74.

(b) *The probability distribution of N*

In order to find the probability distribution of N we use the relationship to the occupancy problem; see Section 5.5.

The event $N \leq n$ is seen to be the same as the event 'in n packages, all r types appear at least once'. This event, in its turn, is the same as the event 'if a die with r faces is thrown n times, all faces appear at least once'. The probability of the latter event is given in expression (3) of Section 5.5. Hence we conclude that

$$P(N \leq n) = \sum_{j=0}^{r} (-1)^j \binom{r}{j} \left(1 - \frac{j}{r}\right)^n \qquad (1)$$

for $n = r, r+1, r+2, \ldots$.

Here is a problem for the reader: Suppose that there are four different prizes in a certain type of food package. How many food packages is it necessary to buy in order to have a probability of at least $\frac{1}{2}$ to obtain all prizes? Using (1), one can show that $n = 7$ (but not $n = 6$) is enough.

Coupon collector's problems are discussed in many papers. A good one is Dawkins (1991), where further references are found.

7.6 Coupon collecting II

We shall discuss a variant of the coupon collector's problem.

In a food package there is a prize belonging to one of two sets A and B. Each set consists of n different types. Again, all types are distributed at random between the packages. A person buys one package at a time until *either* set A *or* set B is complete, whichever comes first. Find the expected value of the number, N, of packages bought. The derivation consists of two parts.

First, consider the number of *different* types of packages that the person buys. When he stops buying, he has obtained, say, R different types. Here R is an rv taking values from n to $2n - 1$. Among these types there are either n of set A or n of set B. The probability function $\{p_r\}$ of R is obtained by drawing without replacement from an urn with n white and n black balls until either all white or all black balls have been drawn. It follows from (7) in Section 7.4 that

$$p_r = 2\binom{r-1}{n-1} \Big/ \binom{2n}{n} \qquad (1)$$

for $r = n, n+1, \ldots, 2n - 1$.

Second, we determine the conditional mean $E(N|R = r)$ of N given the event $R = r$. This is the average number of packages which the person must buy in order to obtain r different out of $2n$ different types. Similarly, as in Subsection (a) of the previous section we find that

$$E(N|R = r) = 2n\left(\frac{1}{2n} + \frac{1}{2n-1} + \cdots + \frac{1}{2n-r+1}\right). \qquad (2)$$

From (1) and (2) we obtain the answer

$$E(N) = \sum_{r=n}^{2n-1} E(N|R = r)p_r.$$

As each factor is known the problem is solved.

For example, if $n = 3$ we find

$$p_3 = \frac{1}{10}; \quad p_4 = \frac{3}{10}; \quad p_5 = \frac{6}{10}$$

and

$$E(N|R = 3) = \frac{37}{10}; \quad E(N|R = 4) = \frac{57}{10}; \quad E(N|R = 5) = \frac{87}{10},$$

and so $E(N) = 73/10$.

Another solution of the problem, leading to a simpler expression for $E(N)$, is given in Section 15.4.

7.7   Ménage II

In Section 5.8 we discussed the classical ménage problem. A modification of this problem will now be considered.

At a dinner for six married couples, the twelve people are placed at random at three tables for four, with the restriction that two women are placed at each table. Find the probability that none of the married couples are placed at the same table.

(a) *First solution*

Call the tables T_1, T_2, T_3. One possible practical way of arranging the seating is the following:

1. Place the women at random at the tables and number their husbands in the following special manner: If a woman is placed at table T_i, give her husband the number i, where $i = 1, 2, 3$. In this way the husbands get personal numbers 1 1 2 2 3 3.

2. Prepare 6 slips of paper marked 1 1 2 2 3 3 and distribute them at random among the men. If a man receives a slip marked j, he is placed at table T_j, where $j = 1, 2, 3$.

We are interested in the event that all men receive a slip number different from the personal number. One favourable case is the following:

Personal numbers	1	1	2	2	3	3
Slip numbers	2	2	3	3	1	1

We now keep the first line fixed and permute the second in all possible ways. There are $6!/2!\,2!\,2! = 90$ possible cases and the following 10 favourable cases:

2	2	3	3	1	1
3	3	1	1	2	2
2	3	1	3	1	2
2	3	1	3	2	1
2	3	3	1	1	2
2	3	3	1	2	1
3	2	1	3	1	2
3	2	1	3	2	1
3	2	3	1	1	2
3	2	3	1	2	1

As a consequence, the probability we seek is equal to $10/90 = 1/9$.

(b) *Second solution*

Call the tables T_1, T_2, T_3 as before. Number the men and women in the couples from 1 to 6. Without loss of generality we may assume that the women 1, 2 are placed at T_1, while 3, 4 are placed at T_2 and 5, 6 at T_3.

Let us divide the event H in which we are interested into two disjoint events H_1 and H_2 as follows:

H_1 : The spouses of the men who happen to be placed at T_1 sit at the same table, that is, either both at T_2 or both at T_3. This means that either the men 3, 4 or the men 5, 6 are placed at T_1. Furthermore, the other pair of men should be placed at the table allotted to the spouses of the men placed at T_1.

H_2 : The spouses of the men placed at T_1 sit at different tables, that is, one at T_2 and one at T_3. This means that either the men 3, 5 or 3, 6 or 4, 5 or 4, 6 are placed at T_1. Furthermore, the two remaining men should be placed at tables not allotted to their spouses.

Multiplying a chain of conditional probabilities, we find

$$P(H_1) = \frac{4}{6} \cdot \frac{1}{5} \cdot \frac{2}{4} \cdot \frac{1}{3} = \frac{8}{360}.$$

(The probability that one of the men 3–6 is placed at T_1 is 4/6; then the other man in this pair remains to be placed there, which happens with probability 1/5. Furthermore, one of the men in the other pair should sit at any of the two places at the table where the spouses of the men at T_1 sit; this happens with probability 2/4. Finally, the other man in the last pair should sit at the same table, which happens with probability 1/3.) In a similar way we find

$$P(H_2) = \frac{4}{6} \cdot \frac{2}{5} \cdot \frac{2}{4} \cdot \frac{2}{3} = \frac{32}{360}.$$

Hence $P(H) = 8/360 + 32/360 = 1/9$.

In the remaining part of the section, we change the problem by removing the condition that two women are placed at each table.

We seek the probability $P(H)$, where $H = AB$, the event A denoting 'no married couple at Table T_1' and the event B 'no married couples at Tables T_2 and T_3'. Let us write

$$P(H) = P(A)P(B|A). \tag{1}$$

Conditioning on the event 'i women at Table 1' we find after some reflection

$$P(A) = \frac{15}{495} \cdot 1 + \frac{60}{495} \cdot \frac{3}{6} + \frac{90}{495} \cdot \frac{4}{6} \cdot \frac{3}{5} + \frac{60}{495} \cdot \frac{5}{6} \cdot \frac{4}{5} \cdot \frac{3}{4} + \frac{15}{495} \cdot 1 = \frac{16}{33}.$$

Moreover, it is realized after some more pondering that, if A occurs, there are exactly two married couples among the people to be placed at Tables 2 and 3. This leads to the conditional probability

$$P(B|A) = \frac{4}{7} \cdot \frac{3}{5} = \frac{12}{35}.$$

Using (1) we are now able to announce the final result

$$P(H) = \frac{16}{33} \cdot \frac{12}{35} = \frac{64}{385} \approx 0.1662.$$

Other routes may be followed, but the one chosen here is believed to be the shortest. This is, indeed, a difficult, but also very instructive, problem.

7.8   Rencontre II

Consider s decks, each consisting of n cards numbered $1, 2, \ldots, n$. All the ns cards are thoroughly mixed and n of them are chosen at random. Let the cards obtained have numbers $(\pi_1, \ldots, \pi_n)$. Let A_i be the event that there is a rencontre at i so that $\pi_i = i$. Find the probability that k or more rencontres occur and the probability that exactly k rencontres occur.

As the A_i's are exchangeable events we can apply formula (2) in Section 2.4. Hence the probability that at least k rencontres occur is given by

$$P_n(k) = \sum_{i=k}^{n} (-1)^{i-k} \binom{i-1}{i-k} \binom{n}{i} P(A_1 \cdots A_i). \tag{1}$$

Using conditional probabilities we can write

$$P(A_1 \cdots A_i) = P(A_1)P(A_2|A_1)P(A_3|A_1A_2) \cdots P(A_i|A_1 \cdots A_{i-1}). \tag{2}$$

Obviously we have

$$P(A_1) = \frac{s}{ns}; \quad P(A_2|A_1) = \frac{s}{ns-1}; \quad P(A_3|A_1A_2) = \frac{s}{ns-2}$$

and so on. Inserting these probabilities in (2) we obtain

$$P(A_1 \cdots A_i) = \frac{s^i}{ns(ns-1)\cdots(ns-i+1)}. \tag{3}$$

Hence (1) can be written in the form

$$P_n(k) = \sum_{i=k}^{n} (-1)^{i-k} \binom{i-1}{i-k} \binom{n}{i} \frac{s^i}{ns(ns-1)\cdots(ns-i+1)}.$$

To obtain the probability that exactly k rencontres occur we can use formula (1) in Section 2.4:

$$p_n(k) = \binom{n}{k} \sum_{i=k}^{n} (-1)^{i-k} \binom{n-k}{i-k} P(A_1 \cdots A_i). \qquad (4)$$

Inserting (3) we find

$$p_n(k) = \binom{n}{k} \sum_{i=k}^{n} (-1)^{i-k} \binom{n-k}{i-k} \frac{s^i}{ns(ns-1)\cdots(ns-i+1)}. \qquad (5)$$

(Note that this formula holds only for $k \geq 1$; $p_n(0)$ is obtained by subtraction.)

Conditional probabilities were used above to calculate $P(A_1 \ldots A_i)$; alternatively we may use the classical rule for determining probabilities in combinatorial problems.

In the case $s = 1$, which leads to the classical rencontre problem, the formulas are simplified. For example, (5) becomes

$$p_n(k) = \frac{1}{k!}\left[1 - \frac{1}{1!} + \frac{1}{2!} - \cdots + (-1)^{n-k}\frac{1}{(n-k)!}\right]. \qquad (6)$$

Example

Consider s ordinary decks of cards, so $n = 52$. The following table gives numerical values for $p_{52}(k)$ when $k = 0, 1, \ldots, 5$ and $s = 1, 2, \ldots, 5$.

$k\backslash s$	1	2	3	4	5
0	0.3679	0.3661	0.3655	0.3652	0.3650
1	0.3679	0.3697	0.3703	0.3706	0.3707
2	0.1839	0.1848	0.1851	0.1853	0.1854
3	0.0613	0.0610	0.0609	0.0609	0.0608
4	0.0153	0.0150	0.0148	0.0148	0.0147
5	0.0031	0.0029	0.0028	0.0028	0.0028

The table shows, for example, that the probability of 2 rencontres when selecting 52 cards from 3 decks is 0.1851.

Here is an easy problem. Suppose two gamblers A and B each have a deck of cards containing 3 and 4 cards, respectively. Both gamblers mix their decks and draw all cards at random, one at the time. Show that the probability that A obtains at least as many rencontres as B is 23/36.

Here is a more difficult problem. Persons A, B and C each have a deck of cars consisting of 3, 4 and 5 cards, respectively. The cards are numbered 1,2,3; 1,2,3,4; 1,2,3,4,5. Each person mixes his deck and determines the number of rencontres. Show that the three people obtain the same number of rencontres with probability 13/120.

8
Poisson approximation

Approximations abound in probability. In particular, one probability distribution is often approximated by another. In this chapter we will consider some situations where a Poisson distribution, and in some cases a binomial distribution, can be used as an approximation of the true distribution of an rv. It is not possible to give a general theory here, only some examples. The classic Feller (1968) contains many interesting examples on Poisson approximation. In Barbour, Holst and Janson (1992) many new results on such approximations are obtained by the method due to Stein and Chen together with so-called coupling methods.

The basic setup for a Poisson approximation is a random experiment in which events happen with small probabilities. The classical and simplest case is of course when the events are independent, but many interesting situations involve 'weakly' dependent events. Let X be the number of events that happen. For the Poisson distribution the mean and the variance are equal, therefore one might expect that if the mean and the variance of X are approximately equal, the distribution of X could be well approximated by a Poisson distribution; under rather general conditions this is indeed true as is shown in Barbour, Holst and Janson (1992).

8.1   Similar pairs and triplets

(a) *Similar pairs*

Call two people with the same birthday a *similar pair*. Let W be the number of similar pairs when n people are compared two at a time. Then W is an rv assuming one of the values $0, 1, \ldots, \binom{n}{2}$.

The distribution of W is complicated, but its expectation is simple. If there are d days in a year we have

$$E(W) = \frac{1}{d} \binom{n}{2}.$$

Even the proof is simple. Write W as a sum of zero–one rv's I_{ij},

$$W = \sum_{i<j} I_{ij},$$

where I_{ij} is 1 if the ith and the jth people constitute a similar pair, and 0 otherwise. The probability that I_{ij} is 1 is equal to $1/d$, and hence its mean is equal to $1/d$. This leads to the formula given above.

If n is not too large compared to d, the distribution of W can be approximated by a Poisson distribution with mean $\binom{n}{2}/d$. For example, we obtain for $n = 23$ and $d = 365$

$$P(W \geq 1) \approx 1 - e^{-\binom{23}{2}/365} \approx 0.5000.$$

This is a good approximation of the correct value 0.5073; see Section 7.1. (Note that the event $W \geq 1$ is identical with the event that two or more persons have the same birthday.)

(b) *Similar triplets*

Call three people with the same birthday a *similar triplet*. Let W be the number of similar triplets when there are d days and n persons. Set

$$W = \sum_{i<j<k} I_{ijk},$$

where I_{ijk} is 1 if the ith, jth and kth persons have the same birthday, and 0 otherwise. This happens with probability $1/d^2$; hence I_{ijk} has mean $1/d^2$. Since there are $\binom{n}{3}$ terms in W, we obtain by a similar argument as in (a):

$$E(W) = \frac{1}{d^2}\binom{n}{3}.$$

Even in this case we can often use the Poisson distribution as an approximation. The reader is invited to apply this approximation to the cases $d = 365, n = 83$ and $d = 365, n = 84$. He will then find

$$P(W \geq 1) \approx 0.4983; \qquad P(W \geq 1) \approx 0.5109,$$

respectively. Hence, *according to this approximation*, 84 people are required in order to make the probability of at least one similar triplet larger than $\frac{1}{2}$. We shall return to this problem in Sections 9.1 and 15.3.

We also invite the reader to solve the following problem: There are $365 \cdot 24 = 8760$ hours in a year. Show that the probability is about $\frac{1}{2}$ that, among 111 randomly selected people, at least two were born the same day and hour.

Reference: Blom and Holst (1989).

8.2   A Lotto problem

In one variant of the game of *Lotto* a person chooses 7 different numbers among $1, 2, \ldots, 35$ each week. The winning choice is each time given by drawing 7 numbers at random without replacement from the same set of numbers. A person playing this game has found that not all the 35 different numbers occurred in the winning choices for 10 weeks. Because 70 numbers have been drawn, each should occur approximately twice. The player thinks that not all numbers are equally likely to be drawn; perhaps some kind of cheating is going on. Let us therefore compute the probability of getting all 35 numbers within 10 weeks.

The probability of not getting number i in a week is evidently 28/35, thus the probability of not getting it in 10 weeks is $(28/35)^{10} \approx 0.1074$. Let $U_i = 1$ if the number i is not obtained in 10 weeks and $U_i = 0$ otherwise. Then the number of missing numbers X can be written

$$X = U_1 + U_2 + \cdots + U_{35}$$

with mean

$$E(X) = 35 \left(\frac{28}{35} \right)^{10} \approx 3.7581.$$

As the probabilities $P(U_i = 1) \approx 0.1074$ are small and the U's are rather 'weakly' dependent, one could contemplate approximating the distribution of X with Po(3.7581), giving the following approximation of the probability:

$$P(X = 0) \approx e^{-3.7581} = 0.0233.$$

It thus seems rather unlikely to get all the numbers in 10 weeks!

More accurate approximations can be obtained by using suitable binomial distributions instead of the Poisson distribution. Using the inclusion-exclusion formula (6) in Section 2.4, it is possible in this case to calculate the probability $P(X = 0)$ exactly. After some computation, one finds this to be 0.0083.

8.3   Variation distance

Let us consider two integer-valued rv's X and Y with probability functions $P(X = k)$ and $P(Y = k)$ for $k = 0, 1, \ldots$. Their joint distribution is not involved; the rv's need not even be defined on the same probability space. We want to approximate the distribution of one of the rv's with that of the other, and measure the accuracy of the approximation. As a measure, we may use some sort of *distance* between the distributions of X and Y. Here are three examples of how a distance may be defined:

$$d_1 = \max_A |P(X \in A) - P(Y \in A)|, \tag{1}$$

$$d_2 = \tfrac{1}{2} \sum_{k=0}^{\infty} |P(X = k) - P(Y = k)|, \tag{2}$$

$$d_3 = \sum_{k=0}^{\infty} [P(X = k) - P(Y = k)]^2. \tag{3}$$

In (1) A is any possible set of one or more integers.

Example

Suppose that X assumes the values 0, 1, 2 with probabilities 0.50, 0.40, 0.10, respectively, and that Y assumes the same values with probabilities 0.80, 0.15, 0.05.

We begin with d_2 and d_3 and find

$$d_2 = \tfrac{1}{2}\big(|0.50 - 0.80| + |0.40 - 0.15| + |0.10 - 0.05|\big) = 0.30,$$

$$d_3 = [(0.50 - 0.80)^2 + (0.40 - 0.15)^2 + (0.10 - 0.05)^2] = 0.155.$$

To find d_1 by a direct calculation we have to consider all possible sets A. There are $2^3 - 1 = 7$ such sets, namely

$$\{0\}, \{1\}, \{2\}, \{0,1\}, \{0,2\}, \{1,2\}, \{0,1,2\}.$$

Calculating $|P(X \in A) - P(Y \in A)|$ in these cases, we find, respectively,

$$0.30, \ 0.25, \ 0.05, \ 0.05, \ 0.25, \ 0.30, \ 0.$$

The largest of these values is 0.30, and so $d_1 = 0.30$. We notice that $d_1 = d_2$. This is no coincidence, for it is always true! (We will not prove this.) These measures will be used here mainly because they are mathematically tractable. However, d_3 is no bad choice; because of the squaring of the differences, extra weight is given to the large ones, which may be advantageous. Naturally, many other measures can be devised.

As a measure of the accuracy of an approximation in this chapter we use the *variation distance*

$$d_V(X, Y) = \max_A | P(X \in A) - P(Y \in A) |. \tag{4}$$

As indicated in the above example we have the convenient alternative form

$$d_V(X, Y) = \tfrac{1}{2} \sum_{k=0}^{\infty} | P(X = k) - P(Y = k) |. \tag{5}$$

Other forms are possible. It is clear from (4) that $0 \leq d_V \leq 1$ and that $d_V = 0$ if and only if X and Y have exactly the same distribution.

Finally, we mention without proof that if X_1, X_2 are independent rv's and Y_1, Y_2 are independent rv's, then

$$d_V(X_1 + X_2, Y_1 + Y_2) \leq d_V(X_1, Y_1) + d_V(X_2, Y_2); \qquad (6)$$

see for example Durrett (1991 p. 120).

8.4   Poisson–binomial

In Section 2.6, Subsection (b), the Poisson–binomial distribution was introduced, that is, the distribution of a sum

$$X = U_1 + U_2 + \cdots + U_n$$

of independent zero–one rv's, where U_i is 1 and 0 with probabilities p_i and $1 - p_i$, respectively.

(a) *Equal p_i's*

If the p_i's are all equal to p, we have $X \sim \mathrm{Bin}(n, p)$. In the book by Poisson from 1837 (see Section 5.1) it was proved that for a small p the binomial distribution can be approximated by a Poisson distribution having mean np. This is also implicit in the work of de Moivre which appeared about a century earlier. As a measure of the accuracy of the approximation, we may use the variation distance introduced in the previous section. One can show that if $Y \sim \mathrm{Po}(np)$, we have

$$d_V(X, Y) \leq np^2; \qquad (1)$$

see for example Durrett (1991, p. 121). Hence the variation distance is small if p is small enough.

The upper bound in (1) is rather crude. For example, if $n = 100$ and $p = 0.05$ we obtain $d_V(X, Y) \leq 0.2500$. A numerical calculation shows that the true variation distance in this case is 0.0128. However, if $n = 100$ and $p = 0.01$, we obtain $d_V(X, Y) \leq 0.0100$ which is more satisfactory. In this case, the true variation distance is 0.0028.

The reader should be aware that, in these days of pocket calculators and personal computers, there is no longer the same need to approximate a binomial distribution with a Poisson distribution unless n is very large.

(b) *The general case*

For unequal p_i's, a simple formula for the probability function of X is generally not available; the expression given by formulas (2) and (4) in Section 3.5 is often too complicated to be of practical use. Therefore approximations are of interest.

By analogy with the case above, it seems reasonable to try to approximate the Poisson–binomial distribution with a Poisson distribution having the same mean, $\lambda = \sum_{i=1}^{n} p_i$. As a measure of the accuracy we again use the variation distance. It can be proved that with an rv $Y \sim \text{Po}(\lambda)$ we have

$$d_V(X,Y) \leq \sum_{i=1}^{n} p_i^2. \tag{2}$$

An elementary proof of this nice inequality is found, for example, in Durrett (1991, p. 121). For large λ the inequality is considerably sharpened by the following inequality due to Barbour and Hall (1984):

$$d_V(X,Y) \leq \frac{1 - e^{-\lambda}}{\lambda} \sum_{i=1}^{n} p_i^2. \tag{3}$$

See also Barbour, Holst and Janson (1992, p. 34).

Example

Suppose that $n = 10$ and $p_1 = \cdots = p_4 = 0.05$ and $p_5 = \cdots = p_{10} = 0.10$. Then $\lambda = 4 \cdot 0.05 + 6 \cdot 0.10 = 0.8$ and (3) gives the inequality

$$d_V(X,Y) \leq \frac{1 - e^{-0.8}}{0.8}(4 \cdot 0.05^2 + 6 \cdot 0.10^2) < 0.05.$$

Hence, if the rv X with the above Poisson–binomial distribution is approximated by an rv Y with a Poisson distribution $\text{Po}(0.8)$, the difference $|P(X \in A) - P(Y \in A)|$ is always less than 0.05.

(c) *Approximation with binomial distribution*

Though somewhat outside the scope of this chapter, we want to mention that it is sometimes advantageous to approximate the rv X with a binomial rv $Y \sim \text{Bin}(m, p)$. We then determine m and p so that X and Y have, approximately, the same means and variances. This leads to the equations

$$mp = \sum_{i=1}^{n} p_i; \qquad mp(1 - p) = \sum_{i=1}^{n} p_i(1 - p_i).$$

Since we want m to be an integer, only approximate equality can generally be attained.

In the example above we get the equations

$$mp = 0.8; \quad mp(1 - p) = 0.8 - 0.07,$$

which gives the approximation Bin(9, 0.089). The value of $P(X = 0)$ to four decimal places is 0.4329; the Poisson approximation gives 0.4493 and the binomial approximation 0.4322.

8.5   Rencontre III

We have already discussed the rencontre problem twice, in Section 5.4 and in Section 7.8. We have determined the probability that a random permutation $(\pi_1, \pi_2, \ldots, \pi_n)$ of the set of numbers $\{1, 2, \ldots, n\}$ has no fixed point; that is, $\pi_i \neq i$ for $i = 1, \ldots, n$.

Now let X be the number of fixed points in the random permutation. We are interested in the mean and variance of X, and in its distribution.

(a) *Mean and variance*

We can represent X as a sum of zero–one rv's

$$X = U_1 + U_2 + \cdots + U_n \tag{1}$$

where U_i is 1 if i is a fixed point and 0 otherwise. It is seen that

$$P(U_i = 1) = \frac{1}{n}$$

and, for $i \neq j$,

$$P(U_i = U_j = 1) = \frac{1}{n(n-1)}.$$

It follows that

$$E(U_i) = \frac{1}{n}; \quad E(U_i U_j) = \frac{1}{n(n-1)},$$

and therefore

$$Var(U_i) = \frac{1}{n} - \frac{1}{n^2}; \quad Cov(U_i, U_j) = \frac{1}{n(n-1)} - \frac{1}{n^2} = \frac{1}{n^2(n-1)}.$$

Applying these results to (1) we immediately obtain

$$E(X) = 1,$$

and after some manipulations,

$$Var(X) = 1.$$

(b) *Distribution*

Since the mean and variance of X are equal, and the same as for an rv $Y \sim$ Po(1), one would expect that its distribution is close to Po(1). Formula (6) in Section 7.8 gives an explicit expression of the probability function:

$$P(X = k) = \frac{1}{k!} \sum_{j=0}^{n-k} \frac{(-1)^j}{j!}.$$

Letting $n \to \infty$ while keeping k fixed we find from this formula that

$$P(X = k) \to \frac{1}{k!} \sum_{j=0}^{\infty} \frac{(-1)^j}{j!} = \frac{e^{-1}}{k!},$$

that is, Po(1) is the limit distribution of X as $n \to \infty$. Furthermore, one can show that with $Y \sim$ Po(1),

$$d_V(X, Y) \approx \frac{2^n}{(n+1)!} \quad \text{as} \quad n \to \infty,$$

indicating that Po(1) gives very accurate approximation. This can also be confirmed by numerical computations. For example, if $n = 10$, we have $d_V(X, Y) \approx 2.6 \cdot 10^{-5}$.

For the rencontre problem discussed in Section 7.8 the reader is invited to prove that $E(X) = 1$ and $Var(X) = 1 - (s-1)/(n-1)$. When n is large and s is moderate one can see with the above arguments that it is reasonable to approximate the distribution of X with Po(1).

8.6 Ménage III

We continue the discussion of the modified ménage problem in Section 7.7. Let us repeat the problem:

Six married couples are randomly placed at three tables for four with the restriction that two men and two women are placed at each table. We suppose that the two men and the two women take alternating seats, but are otherwise randomly seated. Let X be the number of married couples sitting at the same table.

Introduce zero–one rv's $U_1, \ldots, U_{12}$ where $U_i = 1$ if person i has her/his married partner at the right side, and $U_i = 0$ otherwise. We then have

$$X = U_1 + \cdots + U_{12}.$$

We can see that

$$P(U_i = 1) = \frac{1}{6},$$

and hence the expectation of X is given by

$$E(X) = 12 \cdot \frac{1}{6} = 2.$$

It would be interesting to know more about X. What can be said about its distribution? An exact answer can be obtained by making a total enumeration of all the possible seating arrangements; for example, one finds that $P(X = 0) = 1/9$ (see Section 7.7). Here we give three approximate answers:

(a) *First approximation*

Forgetting that the U_i's are dependent, we get the approximation

$$P(X = 0) = P(U_1 = 0, \ldots, U_{12} = 0) \approx \left(\frac{5}{6}\right)^{12} \approx 0.1122.$$

This is a good approximation of the correct value $1/9 \approx 0.1111$. By the same 'forgetfulness' we may approximate the distribution of X by $\text{Bin}(12, 1/6)$.

(b) *Second approximation*

Let $U_i' = 1$, if woman i has her husband on either side and $U_i' = 0$ otherwise. Then we have the representation

$$X = U_1' + \cdots + U_6',$$

where $P(U_i' = 1) = 1/3$. Approximating as in (a) we obtain

$$P(X = 0) \approx \left(1 - \frac{1}{3}\right)^6 \approx 0.0878,$$

corresponding to the approximate distribution $\text{Bin}(6, 1/3)$.

(c) *Third approximation*

Let us use a Poisson distribution with mean $E(X) = 2$ as an approximation.
We get

$$P(X = 0) \approx e^{-2} \approx 0.1353.$$

Of these approximations the first is the best. However, we know that
the correct answer is 1/9, so what? Can we learn anything from the above
considerations?

Suppose, more generally, that $2n$ couples are seated at n tables in
such a way that two men and two women sit at each table. Now $P(X = 0)$
is not so easy to find. However, the expectation can be found as above:
$E(X) = 2$.

We see that the approximations in (a), (b), (c) become $\mathrm{Bin}(4n, 1/2n)$,
$\mathrm{Bin}(2n, 1/n)$ and $\mathrm{Po}(2)$, respectively. But which is the best? We mention
without proof that, for large n, $\mathrm{Bin}(4n, 1/2n)$ has an error (measured by
the variation distance) of order $1/n^2$, while $\mathrm{Bin}(2n, 1/n)$ and $\mathrm{Po}(2)$ have
errors of order $1/n$. Hence, judged to this measure, the first approximation
is the best.

8.7   Occupancy II

This is a continuation of Section 5.5. In an experiment with d equally likely
outcomes, n independent trials are made. The vector of frequencies of the
outcomes $(X_1, \ldots, X_d)$ has the multinomial distribution

$$P(X_1 = i_1, \ldots, X_d = i_d) = \frac{n!}{i_1! \cdots i_d!} \left(\frac{1}{d}\right)^n,$$

where $\sum_{j=1}^{d} i_j = n$; see Section 5.1, Subsection (b). We will discuss two
separate problems connected with this situation when n and d are large.

(a) *Classical occupancy*

When n is much larger than d, we may expect that most outcomes occur
at least once. Hence the number Y of nonoccurring outcomes is probably
small. We will show that Y then has, approximately, a Poisson distribution.
Write

$$Y = U_1 + U_2 + \cdots + U_d$$

where U_i is 1 if outcome i does not occur and 0 otherwise. We have

$$P(U_i = 1) = \left(1 - \frac{1}{d}\right)^n,$$

and so
$$E(Y) = d\left(1 - \frac{1}{d}\right)^n.$$

Now let n and d go to infinity in such a way that

$$d\left(1 - \frac{1}{d}\right)^n \to \lambda,$$

where λ is given in advance. Using (1) in Section 5.5, it can then be shown that
$$P(Y = k) \to \frac{\lambda^k}{k!} e^{-\lambda}.$$

Thus for a large n and d such that $E(Y)$ is moderate in size it might be reasonable to approximate the distribution of Y with a Poisson distribution with the mean $E(Y)$. (Here n must be larger than d.)

(b) *Another occupancy problem*

Let Z be number of outcomes which occur at least m times. This rv can be written as a sum of zero–one rv's and it can be seen that

$$E(Z) = dP(X_1 \geq m),$$

where $X_1 \sim \text{Bin}(n, 1/d)$. For a small m and a large d and n such that $E(Z)$ is moderate in size, the distribution of Z might be approximated by $\text{Po}(E(Z))$. (Also here n must be larger than d.) An application is given in Section 9.1, Subsection (b).

9
Miscellaneous II

This is a continuation of Chapter 7; see the introduction to that chapter.

9.1 Birthdays and similar triplets

In Section 8.1 it was shown that, approximately, 84 people are needed in order to make the probability larger than $\frac{1}{2}$ for getting at least one triplet of people with the same birthday. In the present section, we shall continue the discussion of similar triplets.

(a) *Exact results*

Let the probability of getting at least one similar triplet be q_n for n people. Assume for generality until further notice that a year has d days. We shall show how this probability can be found accurately. For this purpose we use the multinomial distribution; see Section 5.1, Subsection (b).

Set $p_n = 1 - q_n$. Clearly, p_n is the probability that no similar triplet appears among the n people. We have

$$p_n = \sum_{k=0}^{[n/2]} p_{nk},$$

where p_{nk} is the probability of getting k similar pairs but no similar triplets among n people.

According to the multinomial distribution we have

$$p_{nk} = \sum \frac{n!}{i_1! \cdots i_d!} \left(\frac{1}{d}\right)^n,$$

where the i_j's add up to n; the summation should be performed over all i_j's such that k of them are equal to 2, $n - 2k$ are equal to 1 and the remaining $d - k - (n - 2k) = d - n + k$ are equal to 0. Note that the denominator in the first factor is always equal to 2^k.

We will now determine p_{nk}. First, it is clear that $p_{nk} = 0$ when $n > d + k$. Supposing that $n \leq d + k$, we can choose the k days with frequencies 2 in $\binom{d}{k}$ ways. Furthermore, the $n - 2k$ days with frequencies 1 may be chosen, among the remaining $d - k$ days, in $\binom{d-k}{n-2k}$ ways. It follows that

$$p_{nk} = \binom{d}{k}\binom{d-k}{n-2k}\frac{n!}{2^k}\left(\frac{1}{d}\right)^n.$$

We can use the above expression for determining the probabilities p_{nk}, but it is more convenient to use the recursive formula

$$p_{nk} = \frac{(n-2k+2)(n-2k+1)}{2k(d-n+k)}p_{n,k-1},$$

for $k = c+1, c+2, \ldots$, where $c = \max(0, n-d)$. When the quantities p_{nk} have been computed, they are inserted in the expression for p_n.

In this way it is found that for $d = 365$

$$p_{87} \approx 0.5005; \quad p_{88} \approx 0.4889; \quad q_{87} \approx 0.4995; \quad q_{88} \approx 0.5111.$$

Hence 88 people are needed in order to make the probability of getting at least one triplet larger than $\frac{1}{2}$.

(b) Approximate result

We shall display the splendour of Poisson approximation by computing the probability of at least one triplet when 88 people are compared.

As already said, the frequencies $X_1, \ldots, X_{365}$ of birthdays of 88 people are distributed according to a multinomial distribution. Let Z be the number of days with a frequency of at least 3. We want $P(Z \geq 1)$.

Now apply the Poisson approximation in Subsection (b) of Section 8.7. Since $X_1 \sim \mathrm{Bin}(88, 1/365)$, we have

$$E(Z) = 365P(X_1 \geq 3)$$

$$= 365\left[1 - \sum_{k=0}^{2}\binom{88}{k}\left(\frac{1}{365}\right)^k\left(\frac{364}{365}\right)^{88-k}\right] \approx 0.6923.$$

Hence the distribution of Z is approximated by $\mathrm{Po}(0.6923)$, giving

$$P(Z \geq 1) \approx 1 - e^{-0.6923} \approx 0.4996,$$

as an approximation of the correct value 0.5111.

9.2 Comparison of random numbers

Let us compare the two random numbers

$$1\,2\,0\,1\,6\,3\,3\,1\,7$$

$$0\,5\,4\,6\,6\,3\,3\,1\,2$$

We observe that 6 3 3 1 occurs in both numbers, in the same positions. We are interested in such coincidences. Note that if we look for 3-digit numbers, there are two coincidences, since 6 3 3 and 3 3 1 both occur in the same positions.

Let us pose the following general problem. Select two random numbers of the same length n taken from the set $\{1, 2, \ldots, m\}$. If the same k-digit number occurs in the same positions in both numbers, a *coincidence* is said to have taken place. We wish to determine the mean and variance of the number X of coincidences.

(a) *Expectation*

Set

$$X = U_1 + U_2 + \cdots + U_{n-k+1}, \tag{1}$$

where U_i is 1 if a coincidence occurs at positions $i, i+1, \ldots, i+k-1$ and 0 otherwise. Set $p = 1/m$. Each U_i is 1 with probability p^k and 0 with probability $1 - p^k$, and hence has expectation p^k. Applying this to (1) we have the simple result

$$E(X) = (n - k + 1)p^k. \tag{2}$$

(b) *Variance*

Assume that $n \geq 2k - 1$. Taking the variance of X in (1) we obtain

$$Var(X) = A + 2B, \tag{3}$$

where

$$A = \sum_{i=1}^{n-k+1} Var(U_i), \tag{4}$$

$$B = \sum_{i=1}^{n-k} Cov(U_i, U_{i+1}) + \sum_{i=1}^{n-k-1} Cov(U_i, U_{i+2}) + \cdots$$

$$+ \sum_{i=1}^{n-2k+2} Cov(U_i, U_{i+k-1}). \tag{5}$$

Note that all covariances $Cov(U_i, U_j)$ such that $j - i \geq k$ are zero, in view of the fact that two U's at such a distance are independent. Since the variances in (4) are equal as well as the covariances in each sum in (5), A and B can be written in the form

$$A = (n - k + 1)Var(U_1), \tag{6}$$

$$B = \sum_{j=1}^{k-1} (n - k - j + 1)Cov(U_1, U_{j+1}). \tag{7}$$

We shall determine the variances and covariances appearing in (6) and (7). Since U_1^2 and U_1 have the same expectation p^k, we find

$$Var(U_1) = E(U_1^2) - [E(U_1)]^2 = p^k - p^{2k}. \tag{8}$$

The covariances are somewhat more complicated. We have

$$Cov(U_1, U_{j+1}) = E(U_1 U_{j+1}) - E(U_1)E(U_{j+1}) = E(U_1 U_{j+1}) - p^{2k},$$

where $1 \leq j \leq k - 1$. The product $U_1 U_{j+1}$ is 0 unless both U_1 and U_{j+1} are 1. This happens if the two random numbers contain the same two digits in position 1, and also in position 2, ..., position $j + k$; the probability of this event is p^{j+k}. As a consequence, we obtain

$$Cov(U_1, U_{j+1}) = p^{j+k} - p^{2k}. \tag{9}$$

Inserting (8) and (9) into (6) and (7) we obtain the final result: The variance is given by (3), where

$$A = (n - k + 1)(p^k - p^{2k}),$$

$$B = \sum_{j=1}^{k-1} (n - k - j + 1)p^k(p^j - p^k).$$

Here is a problem to solve. Let X be the number of coincidences of length $k = 3$ in two random decimal numbers of length $n = 1000$. Show that X has mean ≈ 1.0 and variance ≈ 1.2.

9.3   Grouping by random division

Select three numbers T_1, T_2, T_3 at random without replacement from the sequence $0, 1, \ldots, 9$. The remaining seven numbers are thereby divided into one, two, three or four groups. Find the expectation μ of the number of groups.

The ten integers can of course be replaced by ten objects lying in a row, of which we select three.

We give two solutions; the first one containing a wonderful application of zero–one rv's, the second an application of exchangeability.

(a) *First solution*

Look at the sequence $0, 1, \ldots, 9$ from left to right. Set $U_k = 1$ if a group begins at position k, and $U_k = 0$ otherwise.

Example:

$$
\begin{array}{cccccccccc}
0 & 1 & 2 & 3 & 4 & 5 & 6 & 7 & 8 & 9 \\
\cdot & T_2 & T_1 & \cdot & \cdot & \cdot & \cdot & \cdot & T_3 & \cdot
\end{array}
$$

Here we have $U_0 = 1$, $U_3 = 1$, $U_9 = 1$, and the remaining U's are 0.

Set $p_k = P(U_k = 1)$. A group begins at 0 if neither T_1, T_2, T_3 is 0, which happens with probability $1 - 3/10 = 7/10$; that is, we have $p_0 = 7/10$. A group begins at 1 if one of T_1, T_2, T_3 is 0 and none is 1, which happens with probability $(1 - 3/10)(1/3) = 7/30$; that is, $p_1 = 7/30$. The remaining p's are also $7/30$.

Let Y be the number of groups. Clearly, we have $Y = U_0 + \cdots + U_9$ and so

$$
\mu = E(U_0) + E(U_1) + \cdots + E(U_9) = \frac{7}{10} + 9 \cdot \frac{7}{30} = \frac{14}{5}.
$$

(b) *Second solution*

Let $N_1 < N_2 < N_3$ be the numbers drawn, ordered with respect to magnitude. Introduce the rv's

$$
X_1 = N_1; \quad X_2 = N_2 - N_1 - 1; \quad X_3 = N_3 - N_2 - 1; \quad X_4 = 9 - N_3.
$$

Take the same example as in (a):

$$
\begin{array}{cccccccccc}
0 & 1 & 2 & 3 & 4 & 5 & 6 & 7 & 8 & 9 \\
\cdot & N_1 & N_2 & \cdot & \cdot & \cdot & \cdot & \cdot & N_3 & \cdot
\end{array}
$$

In this example we have $X_1 = 1$, $X_2 = 0$, $X_3 = 5$, $X_4 = 1$. The number Y of groups is equal to the number of X's that are 1 or more. Therefore set

$$
Y = W_1 + W_2 + W_3 + W_4,
$$

where W_i is 1 if $X_i \geq 1$ and W_i is 0 otherwise. It can be shown that the X's are exchangeable and that each X_i is 0 with probability $3/10$. As a result, W_i is 1 with probability $7/10$ and

$$
\mu = E(Y) = 4 \cdot \frac{7}{10} = \frac{14}{5}
$$

as we found before.

Here is an exercise. Consider a row of m objects. Select n of the objects at random. The remaining $n - m$ objects are thereby divided into groups. Show that the expected number of groups is $(n + 1)(1 - n/m)$. This problem indicates how expected numbers of combinatorial runs can be found in a simple way; compare Section 6.1.

9.4   Records I

Records are often encountered in everyday life. National and world records are set in sports. In the newspapers, headlines like 'the warmest January of the twentieth century' often appear. In this section we shall discuss some probability aspects of records, using a simple model. It should be stressed that this model is too simple for many applications and must be used with care.

Example 1. Successive observations

Perform independent observations on a continuous rv. For example, collect measurements of some unknown quantity. How often does it happen that a record is obtained, that is, a value larger than all the preceding ones?

Suppose that the first eight values are

1	2	3	4	5	6	7	8
1.03	1.01	1.15	1.24	0.99	1.19	1.17	1.25

The first value is always called a record. There are four records in this sequence altogether, nos. 1, 3, 4, and 8. It is often interesting to study also the size of a record, but this is outside the scope of this section; we are here only interested in the position of the records.

Example 2. Card file

Individuals are ordered alphabetically in a card file. First, a single card is placed in the file, then a second card at its proper place, and so on. A record occurs when a new individual is placed first.

For example, consider the following cards placed in the given order in the file:

Card no.	Name
1	Sandell (S)
2	Tukey (T)
3	Blom (B)
4	Holst (H)

Write the names in a 'pyramid'; see Figure 1. There are two records, which are encircled.

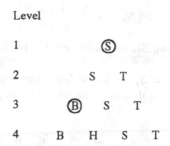

Fig. 1. Records in a card file.

These two examples contain special cases of the following general model: Elements are ordered in a sequence of increasing length according to some rule leading to exchangeability ; see Section 2.2. At each step, the element is inserted at its proper place among those already ordered. If an element comes first, that is, if its *rank* is 1, there is a record. We are interested in the event that a new element receives rank 1. This event is called a *record*. We will study the *position* of the record, that is, the number of elements drawn when the record is attained. We may observe the first record, the second record, and so on. The position of the rth record is denoted by N_r.

We shall study the distribution of the rv N_r. Since we always have $N_1 = 1$, the first interesting case is $r = 2$.

(a) *The second record*

We shall derive the probability function of the position N_2 of the second record. This is very easy in view of the fact that the event $N_2 > k$ (that is, the second record has not yet occurred when k elements have been ordered successively) occurs if and only if the first element has rank 1 among the k elements. This happens with probability $1/k$. Hence we have the simple result

$$P(N_2 > k) = \frac{1}{k} \qquad (1)$$

for $k = 1, 2, \dots$. The expectation of N_2 is infinite, which is remarkable: The second record occurs sooner or later, but, on the average, it takes an infinite amount of time.

As (1) shows, the distribution has a heavy tail to the right. For example, we have $P(N_2 > 100) = 1/100$. As a consequence, one may sometimes have to collect more than 100 elements and order them before the second record occurs.

(b) *The rth record*

In the next section we will show that, for any given r, the rv N_r has the probability function

$$P(N_r = n) = \frac{1}{n!}\begin{bmatrix} n-1 \\ r-1 \end{bmatrix} \tag{2}$$

for $n = r, r+1, \ldots$, using Stirling numbers $\begin{bmatrix} n \\ r \end{bmatrix}$ of the first kind; see Sections 6.2 and 6.3.

For example, if $r = 3$, we know from the second column in Table 1 of Section 6.3 that the probability function of the position N_3 of the third record is given by

n	3	4	5	6	...
$P(N_3 = n)$	$\frac{1}{3!}$	$\frac{3}{4!}$	$\frac{11}{5!}$	$\frac{50}{6!}$	...

We give some references: Glick (1978), Goldie and Rogers (1984), Nevzorov (1987), Nagaraja (1988), Blom, Thorburn and Vessey (1990). Glick's paper contains an excellent introduction to the subject.

9.5   Records II

We use the model introduced in the previous section. Let Y_n be the number of records when n elements are ordered successively. Set

$$Y_n = U_1 + U_2 + \cdots + U_n, \tag{1}$$

where U_i is a zero–one rv, which is 1 if there is a record when element i is inserted among elements $1, 2, \ldots, i-1$ (that is, if element i has rank 1 among elements $1, 2, \ldots, i$), and 0 otherwise. Because of the exchangeability, each of the i first elements has rank 1 with the same probability $1/i$; that is, the probability is $1/i$ that element i has this rank. It follows that the rv U_i has expectation $1/i$ and so, in view of (1), we obtain

$$E(Y_n) = 1 + \frac{1}{2} + \cdots + \frac{1}{n}. \tag{2}$$

For a large n we have $E(Y_n) \approx \ln n + \gamma$, where γ is Euler's constant; see 'Symbols and formulas' at the beginning of the book. Thus the mean number of records increases very slowly with n. This is natural, for as time passes, it becomes more and more difficult to beat the old record.

We may sometimes be interested not only in the mean but in the whole distribution of the number, Y_n, of records. It is derived in the following way:

First, we infer from what has been said about the zero–one rv U_i that $U_i \sim \text{Bin}(1, 1/i)$. Second, we shall prove that the U's are independent. To this end, consider the moment when i elements have been ordered. The rv's $U_1, \ldots, U_{i-1}$ are determined by the order of the elements $1, 2, \ldots, i-1$, but this internal order does not affect the event 'element i has rank 1 compared to these elements'. Hence the U's are independent.

It follows from these considerations that Y_n has a Poisson–binomial distribution; see Section 2.6. (It was proved in Section 6.2 that the number of cycles in a random permutation has exactly the same distribution.) The probability function of Y_n is given by

$$P(Y_n = k) = \frac{1}{n!} \begin{bmatrix} n \\ k \end{bmatrix} \tag{3}$$

for $k = 1, 2, \ldots, n$, using Stirling numbers of the first kind; see Sections 6.2 and 6.3.

For example, using Table 1 in Section 6.3, we get the following probability function for the number of records for $n = 5$:

k	1	2	3	4	5
$P(Y_5 = k)$	$\frac{24}{5!}$	$\frac{50}{5!}$	$\frac{35}{5!}$	$\frac{10}{5!}$	$\frac{1}{5!}$

Finally we will derive the probability function for the waiting time N_r until the rth record occurs. The rth record occurs at the nth position if and only if there are $r - 1$ records in the first $n - 1$ positions and a record at the nth position, that is,

$$N_r = n \quad \text{if and only if} \quad Y_{n-1} = r - 1 \quad \text{and} \quad U_n = 1.$$

As the U's are independent we have

$$P(N_r = n) = P(Y_{n-1} = r - 1)P(U_n = 1).$$

Hence we get from (3)

$$P(N_r = n) = \frac{\begin{bmatrix} n-1 \\ r-1 \end{bmatrix}}{(n-1)!} \cdot \frac{1}{n}$$

for $n = r, r + 1, \ldots$, which proves (2) in the previous section.

9.6   A modification of blackjack

We will consider a modification of the card game 'Blackjack' for two players.

Players A and B draw balls with replacement from an urn containing m balls marked $1, 2, \ldots, m$. The aim of the play is to acquire a ball sum as close to m as possible, without exceeding this number.

First, player A draws a ball and obtains, say, ball i. He may then act in two ways:

Decision 1. A is satisfied with i.

Decision 2. A draws another ball j and so obtains a ball sum $i + j$.

Denote the total number attained by A after one of these decisions by s. If $s > m$, player A loses the game; if $s \leq m$ the game continues as follows:

Player B, who is informed about the value s, draws a ball and obtains, say, ball k. There are two alternatives:

1. $s < k \leq m$. A loses the game.

2. $k \leq s$. Player B draws another ball and obtains, say, ball l. If $k + l \leq s$ or $k + l > m$, player A wins the game; otherwise B wins. (Here we have used the rule that, in the case of a tie, A wins the game.)

We shall now find out when Decision 2 is favourable for A. Clearly, this depends upon the number i.

Let p_i be the probability that A wins the game, given that the first ball has the number i and that he stops after this ball (Decision 1). Similarly, let P_i be the probability that he wins the game, given that the first ball has number i and that he draws another ball (Decision 2).

For any given i, it is favourable for A to draw another ball if $P_i > p_i$. Therefore we have to compare these two quantities for all values of i.

(a) *Calculation of p_i*

If A is satisfied with i, we have $s = i$ and, as seen from alternative 2, A wins if the following events H_1 and H_2 both occur:

$H_1 : k \leq i$
$H_2 : k + l \leq i$ or $k + l > m$.

We shall determine $p_i = P(H_1 H_2)$. We can see that

$$1 - p_i = P(H_1^*) + P(H_1 H_2^*). \tag{1}$$

Remembering that i is a fixed number, we obtain

$$P(H_1^*) = P(k > i) = \frac{m - i}{m}$$

and

$$P(H_1 H_2^*) = P(k \leq i < k + l \leq m)$$

$$= \sum_{v=1}^{i} P(k = v) P(i - v < l \leq m - v | k = v)$$

$$= \sum_{v=1}^{i} \frac{1}{m} \cdot \frac{m - v - (i - v)}{m} = \frac{i(m - i)}{m^2}.$$

Inserting this in (1) we obtain

$$p_i = \left(\frac{i}{m} \right)^2. \tag{2}$$

(b) *Calculation of P_i*

This is the probability that A wins if he obtains ball i in the first draw and decides to draw another ball. His sum is then $i + j$.

We now use what we found in (a). If $j = 1$, the sum is $i + 1$ and A wins with probability p_{i+1}; if $j = 2$, A wins with probability p_{i+2}, and so on. Hence, conditioning on the outcome of A's second draw, we find

$$P_i = \frac{1}{m} \sum_{j=1}^{m-i} p_{i+j} = \frac{1}{m^3} \sum_{j=1}^{m-i} (i + j)^2.$$

Using the well-known formula

$$1^2 + 2^2 + \cdots + n^2 = \frac{1}{6} n(n + 1)(2n + 1)$$

and making some reductions, we obtain

$$P_i = \frac{1}{6m^3} [m(m + 1)(2m + 1) - i(i + 1)(2i + 1)]. \tag{3}$$

We now have to calculate p_i and P_i for all values of i and compare them. For any i such that $p_i \leq P_i$, another ball should be selected, for then A's probability of winning the game is increased (or at least not diminished).

Example

Suppose that there are $m = 20$ balls in the urn. Calculating the quantities

p_i and P_i according to (1) and (2), we obtain the following table:

i	p_i	P_i	i	p_i	P_i
1	0.0025	0.3586	11	0.3025	0.2955
2	0.0100	0.3581	12	0.3600	0.2775
3	0.0225	0.3570	13	0.4225	0.2564
4	0.0400	0.3550	14	0.4900	0.2319
5	0.0625	0.3519	15	0.5625	0.2038
6	0.0900	0.3474	16	0.6400	0.1718
7	0.1225	0.3412	17	0.7225	0.1356
8	0.1600	0.3332	18	0.8100	0.0951
9	0.2025	0.3231	19	0.9025	0.0500
10	0.2500	0.3106	20	1.0000	0.0000

If the first draw gives a ball marked 10 or less, A should select another ball. If the first ball shows 11 or more, A should stop. This strategy seems to be in accordance with what our intuition tells us.

Furthermore, the probability that A wins the game (using the optimal strategy) is given by $p_A \approx 0.4824$, so the game is almost fair.

The corresponding game for three players is discussed in *The American Mathematical Monthly*, Problem E3186 (1989, p. 736).

10
Random walks

Random walks are studied intensively in probability theory. The term was coined by Pólya (1921). In this chapter we shall analyse several walks; for example, walks on the x-axis and in the first quadrant of the (x, y)-plane. Random walk theory has many connections with other parts of probability theory, and is a good field of activity for the probabilist.

Random walks appear in several other chapters of the book, particularly in Chapters 12 and 13.

For further reading we recommend Feller (1968) and Grimmett and Stirzaker (1992). The subject is sometimes studied by sophisticated mathematical methods; see, for example, Spitzer (1964).

10.1   Introduction

We need some terms connected with random walks. The experienced reader is recommended to skip this section and consult it when needed.

A random walk is either *unrestricted* or *stopped*. In the latter case, a *stopping region S* prescribes how the walk ends. In this chapter we mostly discuss stopped walks.

Consider a random walk with two alternative steps at each point of the path. Such a movement can be visualized in at least three different ways:

(a) *Quadrant representation*

The walk takes place in the first quadrant of an (x, y)-system of integer points. It starts from the origin and moves one step to the right with probability p or one step upwards with probability $q = 1 - p$. This movement is repeated step by step until the path reaches the stopping region.

This region can have different shapes (see Figure 1). Generally, it is concave. The walks in Figures 1a and 1b have *closed* stopping regions, whereas the walk in Figure 1c has an *open* stopping region. (A closed stopping region occurs in the discussion of Banach's match box problem; see Section 1.7.)

(b) *Pyramid representation*

The walk takes place in a pyramid as shown in Figure 2. In the simplest case the walk starts from the top and moves downwards either one step to the right with probability p or one step to the left with probability q. Many different stopping regions are possible; for example, a horizontal straight line or an oblique straight line.

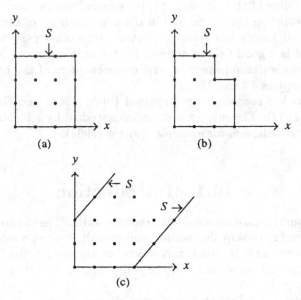

Fig. 1. Quadrant representation. Examples of stopping regions.

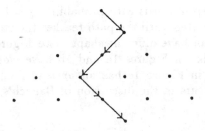

Fig. 2. Random walk. Pyramid representation.

(c) *Line representation*

The walk takes place on the x-axis. One may think of a particle starting at $x = 0$ and moving either right to $x = 1$ with probability p or left to $x = -1$ with probability q. The movement continues until it stops according to some rule. For example, the walk may end when one of the absorbing barriers $x = -a$ and $x = b$ is attained.

Consider a walk with some prescribed stopping rule. Let N be the number of steps until the walk stops. One may be interested in the distribution of N, or only in its expected value. Sometimes the result can be found analytically, sometimes simulation is used. Examples will be given in the following sections.

10.2 Classical random walk II

A particle starts from the origin of the x-axis and moves at times $t = 1, 2, \ldots$ one step to the right or one step to the left according to the following rule: If the particle is at the point $x = i$, it goes right with probability p_i and left with probability $q_i = 1 - p_i$; see Figure 1. The stopping region consists of two barriers $x = -a$ and $x = b$, where a and b are non-negative integers. Each barrier may be either an *absorbing barrier* or a *reflecting barrier*. In the former case the walk stops at the barrier, in the latter case it is reflected. For example, if $x = -a$ is a reflecting barrier and the particle arrives at this point, it moves at the next step with probability 1 to the point $x = -a + 1$.

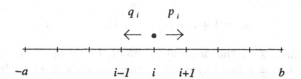

Fig. 1. Random walk with two absorbing barriers.

We now consider the following special situation. The walk starts from the origin as before, but the origin is a right-reflecting barrier ($a = 0$). Hence at the first step, or if the particle returns to the origin, it moves with probability $p_0 = 1$ one step to the right. When the particle reaches the point $x = b$, it is absorbed. We want to find the expected time μ_b from start to absorption. Equivalently, μ_b is the expected number of steps from start to absorption.

Let N_k be the time required for passing from $x = k$ to $x = k + 1$, and $e_k = E(N_k)$ its expected value. Clearly,

$$\mu_n = e_0 + e_1 + \cdots + e_{n-1}$$

is the expected value of the time required for passing from the origin to the point $x = n$.

In order to find e_k we note that

$$N_k = \begin{cases} 1 & \text{with probability } p_k \\ 1 + N_{k-1} + N_k' & \text{with probability } q_k. \end{cases}$$

In the second case, the walk goes to $x = k - 1$ in one step, then back to $x = k$ in N_{k-1} steps and finally to $x = k + 1$ in N_k' steps; the last rv has the same distribution as N_k. Taking expectations and reducing, we obtain

$$e_k = 1 + (e_{k-1} + e_k)q_k.$$

This leads to the recursive relation

$$e_k = \frac{1}{p_k} + \frac{q_k}{p_k}e_{k-1},$$

where $e_0 = 1$. By means of this formula we can find the e's and, by summation, the expected time μ_n.

In the special case when the p's and the q's are $\frac{1}{2}$ for $i = 1, 2, \ldots, b - 1$ we find

$$e_k = 2 + e_{k-1},$$

and so

$$e_k = 2k + 1.$$

This leads to the simple expression

$$\mu_n = 1 + 3 + 5 + \cdots + (2n - 1) = n^2. \tag{1}$$

Hence, returning to the original problem, the mean time to absorption in the point $x = b$ for a symmetric random walk is given by the simple expression

$$E(N) = b^2. \tag{2}$$

Finally, we present a problem to the reader: When $p_i = p$ and $q_i = q$, where $p \neq q$, expression (1) is replaced by

$$\mu_n = cn + d \cdot \frac{1 - (q/p)^n}{1 - q/p},$$

where $c = 1/(p - q)$ and $d = -2q/(p - q)$. Show this.

10.3   One absorbing barrier

A particle performs a symmetric random walk starting from the origin. The walk stops at the absorbing barrier $x = b$.

(a) *First problem*

Show that the particle is absorbed at $x = b$ with probability 1.

Let us first disregard the absorbing barrier and let the particle move freely. Let $P(n)$ be the probability that the particle arrives sooner or later at the point $x = n$, where n is any positive or negative integer. We know, of course, that $P(0) = 1$. If we let the particle walk one step from n, we understand that

$$P(n) = \tfrac{1}{2}P(n+1) + \tfrac{1}{2}P(n-1).$$

Consider the graph of $P(n)$. It follows from the above relation that all points on this graph lie on a straight line. We know that it passes through the point $(0, 1)$. Now suppose that $P(x) < 1$ for a certain value x greater than 0. If this were the case, the straight line would intersect the x-axis and $P(n)$ would be negative for a large enough n. Since this is impossible, we must have $P(n) = 1$ for any $n > 0$. Similarly, it is shown that $P(n) = 1$ for all $n < 0$. This proves the assertion. In particular, we have proved that a symmetric random walk starting from the origin will reach any other given point with probability 1.

(b) *Second problem*

We shall now analyse the particle's arrival at the point $x = 1$ in some detail. Let N be the stopping time until the particle arrives at this point. We shall determine the probability function

$$p_k = P(N = k),$$

where $k = 1, 2, \ldots$.

Let

$$G(s) = E(s^N)$$

be the probability generating function of N. We will determine this function by considering the two possible positions of the particle after making the first jump. If the particle moves to $+1$, we have $N = 1$. If it moves to -1, it must first go to 0, which takes time N', and then to $+1$, which takes time N''; these two rv's are independent and have the same distribution as N. This can be expressed as follows:

$$N = \begin{cases} 1 & \text{with probability } \tfrac{1}{2} \\ 1 + N' + N'' & \text{with probability } \tfrac{1}{2}. \end{cases}$$

This relation leads to a quadratic equation in $G(s)$:

$$G(s) = \tfrac{1}{2}s + \tfrac{1}{2}s\big[G(s)\big]^2.$$

Noting that $G(0) = 0$, we obtain the solution

$$G(s) = \frac{1}{s}\big(1 - \sqrt{1 - s^2}\,\big).$$

Expanding $G(s)$ we find

$$G(s) = \binom{\frac{1}{2}}{1}s - \binom{\frac{1}{2}}{2}s^3 + \binom{\frac{1}{2}}{3}s^5 - \cdots.$$

Here

$$\binom{\frac{1}{2}}{1} = \frac{\frac{1}{2}}{1!} = \frac{1}{2}; \qquad \binom{\frac{1}{2}}{2} = \frac{\frac{1}{2}(\frac{1}{2} - 1)}{2!} = -\frac{1}{8};$$

$$\binom{\frac{1}{2}}{3} = \frac{\frac{1}{2}(\frac{1}{2} - 1)(\frac{1}{2} - 2)}{3!} = \frac{1}{16}$$

and so on, according to the usual rules for forming binomial coefficients (compare Section 7.3). This shows how the probabilities p_k's of the probability function are obtained. The first five p's for an odd n are 1/2, 1/8, 1/16, 5/128, 7/256.

We observe that $\sum p_k = G(1) = 1$, which confirms that absorption occurs sooner or later; compare (a). We also find $E(N) = G'(1) = \infty$. Hence it takes, on the average, an 'infinite' amount of time until absorption takes place.

The reader may perhaps want to study the more general case when the particle goes right or left with probabilities p and $q = 1 - p$, respectively. The probability generating function of N is then given by

$$G(s) = \frac{1}{2qs}\big(1 - \sqrt{1 - 4pqs^2}\,\big).$$

It follows that $G(1)$ is p/q if $p < \tfrac{1}{2}$ and 1 if $p \geq \tfrac{1}{2}$.

The reader is also invited to analyse the situation when the walk stops at any given point $x = a$. (This is easier than it looks at first; if, say, $a = 2$, let the particle first go to 1, and then to 2.) See further Feller (1968, p. 349).

10.4 The irresolute spider

This is a lightweight section written for entertainment.

A spider has produced a cobweb consisting of r 'rays' and n concentric polygons $P_1, P_2, \ldots, P_n$. See Figure 1, where $n = 3, r = 8$. Suddenly, r flies settle down on the outer polygon P_n, one at the end of each ray. The spider walks one step to polygon P_1 along a randomly chosen ray. Thereafter, he either walks back one step or walks to polygon P_2, with the same probability $\frac{1}{2}$. This random walk, forwards or backwards, continues in independent steps. If the spider arrives at the centre, he again chooses a ray at random and continues the walk.

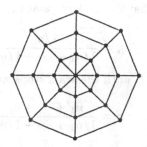

Fig. 1. Cobweb with 8 rays and 3 polygons.

Let N be the number of steps performed by the spider until he catches a fly. The expectation of N is given by the simple expression

$$E(N) = n^2. \tag{1}$$

Note that it does not depend on the number of rays.

For the proof of (1) we use the results of Section 10.2. Suppose that the spider has arrived for the first time at polygon P_k along a nonspecified ray; thus he has not yet visited $P_{k+1}, \ldots, P_n$. Let N_k be the number of additional steps required for reaching P_{k+1}. We can represent N as a sum

$$N = N_0 + N_1 + \cdots + N_{n-1},$$

where $N_0 = 1$. After some consideration we realize that N_k has the same distribution as the number N_k in Section 10.2 has in the symmetric case. (In that section, there was a single ray and reflection at the centre. Now reflection is replaced by random selection among the rays, which is irrelevant.) Hence, expression (2) in Section 10.2 holds even in the present case.

10.5   Stars I

A *star* is a graph with a central vertex and r rays each leading to one
vertex; see Figure 1. A *generalized star* has n_i vertices on the ith ray,
where $i = 1, 2, \ldots, r$. In both cases, a particle performs a random walk
starting from the central vertex. At $t = 1, 2, \ldots$ the particle takes one step
to one of the adjacent vertices, with the same probability. Hence there
are r alternatives to choose between when the particle is at the origin, two
alternatives if the particle is at one of the first $n_i - 1$ vertices on the ith
ray and one if the particle is at the end point of a ray: it then moves to
the $(n_i - 1)$st vertex of that ray.

Let us consider a random walk on a generalized star. The walk stops as
soon as the particle reaches the end of a ray. We shall prove two statements
concerning this walk:

a. The probability q_i that the walk stops at the end of the ith ray is given
   by

   $$q_i = \frac{1/n_i}{1/n_1 + \cdots + 1/n_r}. \tag{1}$$

b. The expected time E until the walk stops at one of the end points is
   given by

   $$E = \frac{n_1 + \cdots + n_r}{1/n_1 + \cdots + 1/n_r}. \tag{2}$$

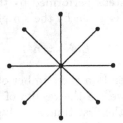

Fig. 1. A star with $r = 8$ rays.

In order to prove (1), we will consider, for any given i, the event A_i
that the walk stops at the end of the ith ray. At the first step from the
central vertex, two situations may occur:

i. The particle arrives at the first vertex on the jth ray, where $j \neq i$.
   From this new starting point, the particle performs a random walk on
   the jth ray until it either reaches the end point of the ray or returns
   to the centre. We now use the solution of Problem 1 in Section 1.5.
   Denoting the new starting point by 0 and taking $a = 1$ and $b = n_j - 1$,

we find that the probability is $1/n_j$ that the particle reaches the end point, and $1 - 1/n_j$ that it reaches the centre. In the former case, the event A_i cannot occur.

ii. The particle arrives at the first vertex on the ith ray. It then reaches the end point of this ray with probability $1/n_i$ and the central vertex with probability $1-1/n_i$. In the former case, the event A_i has occurred.

Noting that at the start each ray is selected with probability $1/r$, we obtain from (i) and (ii) the following equation in the unknown quantity $q_i = P(A_i)$:

$$q_i = \frac{1}{r} \sum_{j \neq i} \left[\frac{1}{n_j} \cdot 0 + \left(1 - \frac{1}{n_j} \right) q_i \right] + \frac{1}{r} \left[\frac{1}{n_i} \cdot 1 + \left(1 - \frac{1}{n_i} \right) q_i \right].$$

Solving with respect to q_i, we obtain (1).

We now prove (2), and need not then discriminate between two cases as in (a):

Let the particle go to the first vertex on the jth ray, where $j = 1, \ldots, r$. Call this point 0. The particle thereafter reaches the end point of this ray with probability $1/n_j$ and the centre with probability $1 - 1/n_j$. We apply the solution of Problem 2 in Section 1.5 with $a = 1, b = n_j - 1$ and conclude that the mean number of steps from the new starting point is $1 \cdot (n_j - 1)$. If the particle reaches the end point, the walk stops, but if it reaches the centre, there are E steps left, on the average.

This reasoning results in the equation

$$E = 1 + \frac{1}{r} \sum_{j=1}^{r} \left[(n_j - 1) + \left(1 - \frac{1}{n_j} \right) E \right].$$

Solving respect to E, we obtain (2).

10.6   Closed stopping region

Consider the quadrant representation of a random walk with closed concave stopping region S [see Section 10.1, Subsection (a)]. The particle starts from the origin and goes right with probability p and up with probability $q = 1 - p$. We shall derive a simple result for the mean $E(N)$ of the number of steps until the walk stops.

Let S^o be the set of points (i, j) inside S. Set U_{ij} equal to 1 if the path passes such a point and equal to 0 otherwise. We may then represent N as a sum

$$N = \sum_{i,j} U_{ij} \tag{1}$$

with summation over S^o. To facilitate the understanding of (1) we give an example.

Example

Assume that the walk stops when it arrives at $x = 3$ or $y = 2$, whichever happens first. The inner points are

$$(0,0), (1,0), (2,0), (0,1), (1,1), (2,1).$$

Suppose that the path happens to be

$$(0,0) \rightarrow (1,0) \rightarrow (1,1) \rightarrow (2,1) \rightarrow (3,1).$$

We then have $N = 4$. On the other hand, we obtain

$$U_{00} = U_{10} = U_{11} = U_{21} = 1; \; U_{20} = U_{01} = 0.$$

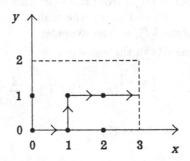

Fig. 1. Closed stopping region.

Hence (1) is satisfied. See Figure 1, where the inner points have been marked with dots.

We now return to the general case and use (1) as follows. According to the binomial distribution, the walk passes the inner point (i, j) with probability

$$\binom{i + j}{i} p^i q^j.$$

Hence this is the probability $P(U_{ij} = 1)$. Since

$$E(N) = \sum E(U_{ij}) = \sum P(U_{ij} = 1),$$

we obtain

$$E(N) = \sum \binom{i+j}{j} p^i q^j \tag{2}$$

with summation over all pairs (i, j) in the set S^o. This is a beautiful application of zero–one rv's.

The formula (2) is convenient for computer calculations. Try it when the stopping region is a quarter of a circle.

It is instructive to apply formula (2) to the coin problem mentioned in connection with Banach's match box problem in Section 1.7. The biased coin is tossed until heads appears r times or tails appears r times, whichever comes first. It follows from (2) that the number N of tosses has expectation

$$E(N) = \sum_{i=0}^{r-1} \sum_{j=0}^{r-1} \binom{i+j}{i} p^i q^j.$$

The coin problem appears in *The American Mathematical Monthly*, Problem E3386 (1990, p. 427 and 1992, p. 272). A proof is asked for that $E(N)$, considered as a function of the probability p, has its maximum when $p = \frac{1}{2}$. Using the above formula, written in the form

$$E(N) = \sum_i \binom{2i}{i} p^i q^i + \sum_{i<j} \binom{i+j}{i} (p^i q^j + p^j q^i),$$

we realize that each term in both sums is largest when $p = \frac{1}{2}$. Therefore, $E(N)$ is then as large as possible.

10.7 The reflection principle

Let us consider a random walk with transition probabilities p and q. We use the pyramid representation of the walk; see Section 10.1. We shall determine the probability f_{2n} that the walk returns to the line of symmetry for the first time after exactly $2n$ steps. For this purpose we shall use the famous *reflection principle*, which has many applications. Number the positions as in Figure 1. Let us start the walk in position $a : m$ and follow some path to position $b : n$, where $n > m$. The number $N_{m,n}(a, b)$ of possible paths between these positions can be found as follows:

Let a path consist of α steps to the right and β steps to the left. We have $\alpha + \beta = n - m$ and $\alpha - \beta = b - a$. Solving for α we obtain

$$\alpha = \tfrac{1}{2}(n - m + b - a). \tag{1}$$

As a result, we have

$$N_{m,n}(a, b) = \binom{n - m}{\alpha}, \tag{2}$$

where α is given by (1).

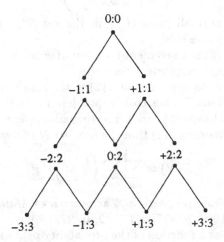

Fig. 1. Numbering of the positions in a pyramid.

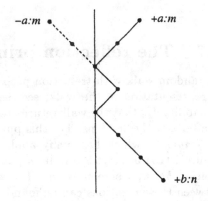

Fig. 2. Reflection of a path in the symmetry line.

Now assume that a and b are positive. We want to determine the number $N^0_{m,n}(a, b)$ of paths that do not touch or intersect the vertical symmetry line. We first determine the number $N^1_{m,n}(a, b)$ of paths with the opposite property; that is, the paths that touch or intersect the symmetry line. For this purpose we use the reflection principle. In order to find all paths with this property we reflect the uppermost part of each such path in the symmetry line and count *all* paths from $-a : m$ to $b : n$; see Figure 2. Apparently, this count will give the wanted result. Hence we have to our satisfaction the simple and elegant expression

$$N^1_{m,n}(a, b) = N_{m,n}(-a, b),$$

and so we find

$$N^0_{m,n}(a, b) = N_{m,n}(a, b) - N^1_{m,n}(a, b) = N_{m,n}(a, b) - N_{m,n}(-a, b). \quad (3)$$

After these preparations we determine f_{2n}. Consider a walk starting from the top of the pyramid, which returns to the line of symmetry for the first time at position $0 : 2n$. The path is divided into three parts:

a. The path first takes one step to the right or one step to the left, say to the right.
b. The path goes from $1 : 1$ to $1 : (2n - 1)$ without touching or crossing the symmetry line.
c. The path goes from $1 : (2n - 1)$ to $0 : 2n$.

The total probability of this alternative is

$$p \cdot N^0_{1,2n-1}(1, 1)p^{n-1}q^{n-1} \cdot q.$$

Adding the 'left' alternative we obtain

$$f_{2n} = 2N^0_{1,2n-1}(1, 1)p^n q^n.$$

An application of (3) shows that

$$N^0_{1,2n-1}(1, 1) = N_{1,2n-1}(1, 1) - N_{1,2n-1}(-1, 1).$$

Using (2) and (1) as well, we find

$$N^0_{1,2n-1}(1, 1) = \binom{2n-2}{n-1} - \binom{2n-2}{n} = \frac{1}{n}\binom{2n-2}{n-1}.$$

This finally leads to

$$f_{2n} = \frac{2}{n}\binom{2n-2}{n-1}p^n q^n,$$

where $n = 1, 2, \ldots$.

The following is a good exercise: Let

$$G(s) = \sum_{n=1}^{\infty} f_{2n}s^{2n}$$

be the generating function of the f's. Prove that

$$G(s) = 1 - \sqrt{1 - 4pqs^2}.$$

It follows that

$$G(1) = 1 - \mid p - q \mid.$$

Therefore, the probability is $\mid p - q \mid$ that the particle never returns to the symmetry line.

References: Feller (1968, p. 273, 313), Grimmett and Stirzaker (1992).

10.8   Ballot problem

In 1887 the *ballot problem* was formulated by J. Bertrand and solved by D. André.

In a ballot, candidate A scores a votes and candidate B scores b votes, where $a > b$. Find the probability P that, throughout the counting, the number of votes for A is always larger than that for B, assuming that all possible orders of voting records are equally probable.

For the solution we use the reflection principle; see the preceding section. (Note that we now use a different numbering of the points in the pyramid.) The successive votes can be represented by a random walk in a pyramid; see Figure 1. The walk starts from 0 : 0 and goes to the right when A wins a vote and to the left when B wins. The walk stops at position $a : (a + b)$.

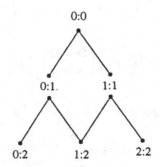

Fig. 1. The ballot problem and random walk.

The number of possible paths from $0 : 0$ to $a : (a + b)$ is equal to

$$\binom{a + b}{a}.$$

All favourable paths must pass from $1 : 1$ to $a : (a + b)$ without touching or crossing the vertical symmetry line. Denote the number of such paths by N_0. Clearly, N_0 is equal to the difference between the number N of all paths from $1 : 1$ to $a : (a + b)$ and the number N_1 of paths between these points that touch or cross the symmetry line. We can see that

$$N = \binom{a + b - 1}{a - 1}.$$

Moreover, in view of the reflection principle, N_1 is equal to the number of possible paths from $0 : 1$ to $a : (a + b)$, and so

$$N_1 = \binom{a + b - 1}{a}.$$

Hence the number of favourable paths is equal to

$$N_0 = N - N_1 = \binom{a + b - 1}{a - 1} - \binom{a + b - 1}{a}.$$

Dividing by $\binom{a+b}{a}$ and reducing, we obtain the nice answer

$$P = \frac{a - b}{a + b}. \tag{1}$$

Takács (1962) has obtained a more general result, which is sometimes called *Takács's theorem*: An urn contains n balls marked $a_1, a_2, \ldots, a_n$, where the a's are non-negative numbers with sum $k \leq n$. Draw all balls at random without replacement. The probability that the sum of the first r numbers drawn is less than r for every $r = 1, 2, \ldots, n$ is equal to $(n - k)/n$.

The solution (1) of the ballot problem can also be used in the following problem: Permute a sequence of n ones and n zeros in all possible ways, and select one of these permutations at random. Represent the permutation as a random walk using the quadrant representation. The probability that the path does not touch or cross the line $x = y$ before it arrives at the point (n, n) is equal to $1/(2n - 1)$. The reader is invited to prove this.

Reference: See the comprehensive article on ballot problems in *Encyclopedia of Statistical Sciences* (1982–1988), Barton and Mallows (1965) and Takács (1989, 1992).

10.9   Range of a random walk

We believe, but are not absolutely sure, that this section contains a new result.

A particle performs a symmetric random walk on the x-axis starting from the origin. At $t = 1, 2, \ldots$ the particle moves one step to the left or one step to the right with the same probability $\frac{1}{2}$. Let

$$R(t) = \{a, a+1, \ldots, b\},$$

where $a \leq 0$ and $b \geq 0$ be the set of points visited up to time t, including the starting point. More briefly, we write

$$R(t) = [a, b]$$

and call $R(t)$ the *range* of the walk at time t.

The behaviour of the range for given t has been studied by Spitzer (1964), and others. We shall study the time when the number of points visited has just increased to a given number i, where i is one of the numbers $1, 2, 3, \ldots$. We call this time T_i, where $T_1 = 0, T_2 = 1$, and denote the range of the walk at this time by $R_i = R(T_i)$. It follows from the rules of the walk that there are i alternatives for R_i, namely

$$A_0 = [-i+1, 0]; \ A_1 = [-i+2, 1]; \ \ldots ; \ A_{i-1} = [0, i-1].$$

At time T_i, the particle is either at the left or at the right end point of the range.

Example

Suppose that, up to time $t = 6$, the path is

$$0 \to 1 \to 0 \to -1 \to 0 \to 1 \to 2.$$

Then $R(6) = \{-1, 0, 1, 2\}$, and so four different points have been visited, Using the condensed notation, we have $R(6) = [-1, 2]$. We have $T_2 = 1, T_3 = 3, T_4 = 6$ and, for example, $R_4 = [-1, 2]$.

Turning to the general case, we shall prove two statements concerning the time T_i and the range R_i.

(a) *The expectation of* T_i

The expectation of T_i is given by

$$E(T_i) = 1 + 2 + \cdots + (i-1) = \tfrac{1}{2}i(i-1). \tag{1}$$

For the proof, we consider the classical symmetric random walk that starts in 0 and has absorbing barriers in $-a$ and b. The expected time until arrival at one of these points is equal to ab; see Problem 2 in Section 1.5. (The same result is, of course, obtained if the starting point is c and the absorbing barriers are $-a+c$ and $b+c$.)

Let us consider the part of the walk from T_i to T_{i+1}. Suppose that at T_i the range is $R_i = [-i+k+1, k]$ and that the particle is at the right end point k. (We might just as well have considered the left end point.) At time T_{i+1} the particle is just outside this range, either at $-i+k$ or at $k+1$. By the classical result just quoted, such a passage takes, on the average, $1 \cdot i$ steps.

Hence we have proved that

$$E(T_{i+1}) - E(T_i) = i,$$

which leads to (1).

(b) *A property of R_i*

As mentioned before, there are i alternatives $A_0, A_1, \ldots, A_{i-1}$ for the range R_i. We shall prove that these alternatives occur with the same probability $1/i$.

To this end, introduce the i pairs of events $(A_0^-, A_0^+), \ldots, (A_{i-1}^-, A_{i-1}^+)$. For example, A_0^- indicates that, at time T_i, the particle is at the left end point of $A_0 = [-i+1, 0]$, while A_0^+ indicates that it is at the right end point.

Consider such a pair (A_k^-, A_k^+) , where

$$A_k = [-i+k+1, k].$$

The event A_k^+ occurs if and only if the following two events happen:

1. The particle visits $-i+k+1$ before k.
2. Thereafter, the particle visits k before $-i+k$.

In view of the classical random walk referred to earlier, the particle visits $-i+k+1$ before k with probability

$$\frac{k}{(i-k-1)+k} = \frac{k}{i-1};$$

see Problem 1 in Section 1.5.

Moreover, starting from $-i+k+1$, the particle visits k before $-i+k$ with probability

$$\frac{1}{(i-k)+k} = \frac{1}{i}.$$

Hence we obtain

$$P(A_k^+) = \frac{k}{i-1} \cdot \frac{1}{i}.$$

Leaving out the details, we find by a similar argument that

$$P(A_k^-) = \frac{i-k-1}{i-1} \cdot \frac{1}{i}.$$

Adding these two probabilities we arrive at the final conclusion that

$$P(A_k) = \frac{1}{i}.$$

11
Urn models

An urn with balls of different colours is one of the probabilist's favourite
toys, which can be used for many serious purposes. A multitude of discrete
phenomena can be modelled by selecting balls from urns, perhaps returning
them or transferring them to other urns. The simplest of these models are
known to any beginner: drawings with or without replacement. In this
chapter, Pólya's and Ehrenfest's famous urn models will be the focus of
our attention, together with some other models.

The best general reference to urn models is Johnson and Kotz (1977).
A wealth of information can be extracted from this book.

11.1   Randomly filled urn

An urn is filled with m balls, each being white with probability p and black
with probability $q = 1 - p$. As a consequence, the number X_m of white
balls in the urn has a binomial distribution $\text{Bin}(m, p)$. We then say that
the urn has been *randomly filled*.

As is well known from previous chapters, we can represent X_m as a
sum

$$X_m = U_1 + U_2 + \cdots + U_m, \tag{1}$$

where U_i is 1 if the ith ball placed in the urn is white and 0 if it is black.

We now draw a sample of n balls from the urn, in two different ways:

(a) *Drawings with replacement*

In this case n can be any number. Let Y_n be the number of white balls in
the sample. Set

$$Y_n = W_1 + W_2 + \cdots + W_n, \tag{2}$$

where W_i is 1 if the ith ball drawn is white and 0 if it is black.

We shall prove that the W's are dependent. For this purpose, first let
the U's in (1) be fixed: k of them are assumed to be 1 and the remaining
$m - k$ equal to 0; hence we have $X_m = k$. Let $E(\cdot | k)$ denote conditional
expectation given that $X_m = k$. We find

$$E(W_i | k) = \frac{k}{m}; \qquad E(W_i^2 | k) = \frac{k}{m},$$

and for $i \neq j$

$$E(W_i W_j | k) = \left(\frac{k}{m}\right)^2.$$

Taking means, noting that $E(X_m) = mp$ and $Var(X_m) = mpq$, we obtain

$$E(W_i) = E(W_i^2) = E\left(\frac{X_m}{m}\right) = p,$$

$$E(W_i W_j) = E\left[\left(\frac{X_m}{m}\right)^2\right] = \frac{1}{m^2}\left\{Var(X_m) + [E(X_m)]^2\right\}$$

$$= \frac{pq}{m} + p^2.$$

It follows that

$$E(W_i W_j) > E(W_i)E(W_j),$$

which proves that the W's are dependent.

(b) *Drawings without replacement*

In this case we must have $n \leq m$. Let Y_n have the same meaning as in (a), and again use representation (2). The W's now have the remarkable property that they are iid and $Bin(1, p)$. The reason is obvious: they constitute a subset of the U's in (1), which themselves are iid and $Bin(1, p)$.

This property has a consequence for quality control. Let the urn represent a batch of m units produced at a factory, and let X_m be the number of defective units. Moreover, let Y_n be the number of defective units in a sample of n units selected at random without replacement from the batch. Then $X_m - Y_n$ is the number of defectives in the non sampled part of the batch. Now assume that the urn is randomly filled. Then the rv's Y_n and $X_m - Y_n$ are independent. This is due to the fact that these two rv's can be represented as sums of n and $m - n$ U's, respectively, which are all independent.

As a result, the sample of n units conveys no information about the quality of the rest of the batch, and thus the inspection is worthless. Remember the condition for the correctness of this statement: the urn is randomly filled or, translated to the language of quality control, the batch is produced under constant conditions with a fixed probability p that a unit is defective.

This consequence for quality control has been known for a long time; see Mood (1943).

11.2 Pólya's model I

In the next three sections we will consider *Pólya's urn model*. *George Pólya* (1887–1985) was a famous mathematician. One can read about him in the Pólya Picture Album (1987). An early paper on Pólya's urn model is Eggenberger and Pólya (1923). Much information about the model is given in the book by Johnson and Kotz (1977).

The model is obtained as follows. An urn contains a white and b black balls. A ball is drawn at random and is returned to the urn together with c balls of the same colour as that drawn. This procedure is performed repeatedly. The total number of balls in the urn will then increase from $a + b$ to $a + b + c$, then to $a + b + 2c$, and so on. A stopping-rule prescribes when to finish the drawings.

Pólya's model may be used as a model for contagion. If a case of disease occurs, the probability of a further case increases. It is perhaps uncertain whether real data would fit the model accurately, but the general idea is of interest.

Now assume that balls are drawn successively according to the model. Let A_i be the event that 'the ith ball drawn is white', where $i = 1, 2, \ldots$. It can be proven that the A's constitute an infinite sequence of exhangeable events; see Section 2.2 for the definition and Feller (1971, p. 229) for a proof.

In practice, we often draw a fixed number, say n, of balls. Denote the number of white balls then obtained by X. We are interested in the distribution of X. The events $A_1, A_2, \ldots, A_n$ constitute a finite sequence of exchangeable events.

Example

Consider an urn with 2 white and 5 black balls. At each drawing, a ball is selected at random and is returned to the urn together with one ball of the same colour as the ball just drawn. Let us draw 4 balls. We have in this case $a = 2$, $b = 5$, $c = 1$, $n = 4$.

Let us write down the probabilities of the sequences that contain 3 A's and 1 A^*:

$$P(A_1 A_2 A_3 A_4^*) = \frac{2}{7} \cdot \frac{3}{8} \cdot \frac{4}{9} \cdot \frac{5}{10} = \frac{1}{42},$$

$$P(A_1 A_2 A_3^* A_4) = \frac{2}{7} \cdot \frac{3}{8} \cdot \frac{5}{9} \cdot \frac{4}{10} = \frac{1}{42},$$

$$P(A_1 A_2^* A_3 A_4) = \frac{2}{7} \cdot \frac{5}{8} \cdot \frac{3}{9} \cdot \frac{4}{10} = \frac{1}{42},$$

$$P(A_1^* A_2 A_3 A_4) = \frac{5}{7} \cdot \frac{2}{8} \cdot \frac{3}{9} \cdot \frac{4}{10} = \frac{1}{42}.$$

Because of the exchangeability, these four probabilities are the same. Thus, the probability of obtaining three white balls in four drawings is given by

$$P(X = 3) = \frac{4}{42} = \frac{2}{21}.$$

In the general case, the probability function of X can be written

$$P(X = k) = \binom{n}{k} \frac{A_k B_{n-k}}{C_n} \tag{1}$$

for $k = 0, 1, \ldots, n$, where $A_0 = B_0 = 1$ and

$$
\begin{aligned}
A_k &= a(a + c)(a + 2c) \cdots [a + (k - 1)c], \\
B_{n-k} &= b(b + c)(b + 2c) \cdots [b + (n - k - 1)c], \\
C_n &= (a + b)(a + b + c)(a + b + 2c) \cdots [a + b + (n - 1)c].
\end{aligned}
$$

Here the factor $\binom{n}{k}$ is the number of orderings resulting in k white and $n - k$ black balls, and $A_k B_{n-k}/C_n$ is the probability of obtaining k white and $n - k$ black balls in some given order. Because of the exchangeability, this probability is the same for all orderings of the k white and $n - k$ black balls.

The probabilities $P(X = k)$ can be calculated recursively. We leave it to the reader to construct such a recursion.

11.3   Pólya's model II

An urn initially contains one white and one black ball. At each drawing, a ball is selected at random and returned to the urn together with one ball of the same colour as the one drawn. The procedure is repeated until n balls have been drawn.

This urn model is obtained by taking $a = 1$, $b = 1$, $c = 1$ in the description of Pólya's urn model in the preceding section. Let X be the number of white balls obtained. We are interested in the distribution of X. It can easily be obtained, for example from the preceding section, but we prefer to use de Finetti's theorem.

Let A_i be the event that 'the ith ball drawn is white', where $i = 1, 2, \ldots, n$. The events A_i are exchangeable. Since they can be extended into an infinite sequence, we are entitled to apply de Finetti's theorem in Section 2.3.

Taking $k = n$ in formula (4) of Section 2.3, we obtain

$$P(X = n) = \int_0^1 p^n f(p)\, dp,$$

which holds for any $n = 1, 2, \ldots$. On the other hand, directly from the model we obtain

$$P(X = n) = \frac{1}{2} \cdot \frac{2}{3} \cdots \frac{n}{n+1} = \frac{1}{n+1}.$$

(The probability is $1/2$ that the first ball is white; if the first was white, it is $2/3$ that also the second is white, and so on.) Thus we have identically

$$\int_0^1 p^n f(p) \, dp = \frac{1}{n+1}.$$

This formula provides us with the moments around the origin of the prior distribution. These moments are well known; they are the moments of the uniform distribution $U(0, 1)$, and no other distribution has these moments. Hence, to find the probability function $P(X = k)$ we use the binomial distribution and then integrate p over the interval $(0, 1)$:

$$P(X = k) = \int_0^1 \binom{n}{k} p^k (1 - p)^{n-k} \, dp.$$

Evaluation of the integral and reduction yields the simple result

$$P(X = k) = \frac{1}{n+1}$$

for $k = 0, 1, \ldots, n$. Hence X has a uniform distribution over the integers $0, 1, \ldots, n$. This is a delicious result.

We leave it to the interested reader to derive this result from (1) in the previous section as well.

11.4 Pólya's model III

As an introduction we shall describe a general random walk with variable transition probabilities. The walk starts from the top $1 : 1$ of a pyramid as illustrated in Figure 1. The figure shows how the positions are numbered. The first step goes with probability p_{11} to position $2 : 2$ and with probability $q_{11} = 1 - p_{11}$ to position $1 : 2$. Generally, it goes from position $i : n$ to $(i+1) : (n+1)$ with probability p_{in} and to $i : (n+1)$ with probability q_{in}. The walk stops at some stopping region S.

We consider Pólya's urn model discussed in the preceding section. At the start, there is one white and one black ball in the urn. A ball is chosen at random and is returned together with one ball of the same colour as the one drawn. This procedure is repeated until S is reached.

Let us now consider the corresponding random walk. At each step the walk goes to the right if a white ball is drawn and to the left if a black ball is drawn. After $n - 1$ steps the walk has reached some position $i : n$ on level n. There are then i white and $n - i + 1$ black balls in the urn.

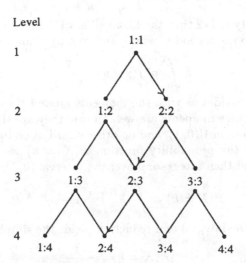

Fig. 1. Random walk in pyramid.

Hence the probability is $i/(n+1)$ that a white ball is drawn next time so that the walk goes right, and $(n-i+1)/(n+1)$ that a black ball is drawn so that the walk goes left. Using the general notation introduced above, we have

$$p_{in} = \frac{i}{n+1}; \qquad q_{in} = \frac{n-i+1}{n+1}.$$

We consider two different stopping regions for the walk.

(a) *Oblique stopping line*

Start as before from $1:1$ and stop when the walk has taken a fixed number, $j-1$, of steps to the right; that is, when $j-1$ white balls have been obtained. It has then arrived at a position $j:k$ on an oblique stopping line. The second index is an rv K taking the value k with probability $(j-1)/[(k-1)k]$, where $k = j, j+1, \ldots$.

To prove this, we first hold p constant. The walk goes first from $1:1$ to $(j-1):(k-1)$ in $k-2$ steps of which $j-2$ are to the right. This happens with probability

$$\binom{k-2}{j-2} p^{j-2}(1-p)^{k-j}.$$

The last step must go to $j : k$, which happens with probability p. Hence we multiply the above binomial probability by p and integrate from 0 to 1 according to the uniform prior; compare the preceding section. Performing the integration we obtain

$$P(K = k) = \int_0^1 \binom{k-2}{j-2} p^{j-1}(1-p)^{k-j}\, dp = \frac{j-1}{(k-1)k},$$

as stated above. Here we have $k = j, j+1, \ldots$.

(b) *Vertical stopping line*

The walk starts at the top of the pyramid and stops at a return to the vertical line of symmetry. This is the first time when equal numbers of white and black balls have been obtained. We shall determine the probability that such a return takes place.

First, we hold p constant in the way we did in (a), and determine the probability that the return takes place after exactly $2n$ steps. For a given p this probability is given by

$$\frac{2}{n}\binom{2n-2}{n-1} p^n(1-p)^n,$$

as found in Section 10.7. Second, integrating from 0 to 1 (remembering that the prior is uniform over this interval), we obtain

$$\int_0^1 \frac{2}{n}\binom{2n-2}{n-1} p^n(1-p)^n\, dp = \frac{1}{(2n-1)(2n+1)}.$$

If N denotes the number of steps until the return, we find

$$P(N \le 2n) = \sum_{j=1}^n \frac{1}{(2j-1)(2j+1)} = \frac{1}{2}\sum_{j=1}^n \left(\frac{1}{2j-1} - \frac{1}{2j+1}\right) = \frac{n}{2n+1}.$$

From this we see that
$$P(N < \infty) = \tfrac{1}{2}.$$

The reader who thought that return to the vertical line is a sure event will be surprised: the particle returns only with probability $\frac{1}{2}$, and, with the same probability, will never return to this line.

Remark

Let us end with a remark showing that the special model treated in this section is more general than might be thought. If position $i : n$ is *given*, the rest of the walk is equivalent to that obtained when drawing balls from an urn with i white and $n-i+1$ black balls according to Pólya's usual rules. Therefore, a general Pólya urn model can be obtained by conditioning in a suitable way in the special one.

Reference: Blom and Holst (1986).

11.5   Ehrenfest's model I

In this section, we combine probability with mathematics. It is a section for professional probabilists, and for undergraduates who want to test their mathematical ability.

Ehrenfest's urn model is known from theoretical physics as a model for the diffusion of molecules between two receptacles. The model can be described as follows, using balls instead of molecules: Two urns U_1 and U_2 contain m balls altogether. At each time-point $t = 1, 2, \ldots$, one of the m balls is drawn at random and is moved from the urn it is taken from to the other urn. Let E_k be the state 'k balls are in urn U_2', where $k = 0, 1, \ldots, m$. Usually, one studies the distribution of the number of balls in each urn after a long time has elapsed; see Section 13.4. In the present section we will consider a less known problem.

Let us assume that at $t = 0$ we are in state E_0: all balls are in U_1. We want to find the expected time μ_n required for the transition from state E_0 to state E_n, where $1 \le n \le m$. As in Section 10.2, set

$$\mu_n = e_0 + e_1 + \cdots + e_{n-1},$$

where e_k is the mean transition time from E_k to E_{k+1}.

In Section 10.2 we derived the recursive relation

$$e_k = \frac{1}{p_k} + \frac{q_k}{p_k} e_{k-1}, \tag{1}$$

with $e_0 = 1$ and $q_k = 1 - p_k$. Here p_k denotes the probability that the number of balls in U_2 increases from k to $k + 1$; this happens if one of the $m-k$ balls in U_1 is selected; hence we have $p_k = (m-k)/m = 1-k/m, q_k = k/m$. This leads to the formula

$$e_k = \frac{m + k e_{k-1}}{m - k},$$

where $k = 1, 2, \ldots, m - 1$.

It is not altogether easy to find an explicit expression for e_k. One possibility is to represent it as an integral

$$e_k = m \int_0^1 x^{m-k-1} (2 - x)^k \, dx.$$

This expression is proved by induction in the recursive relation (1), performing a partial integration. (We omit the details.)

We obtain μ_n by adding over k from 0 to $n-1$. Summing the resulting geometric series in the integrand and performing the substitution $x = 1 - y$ in the integral, we finally find

$$\mu_n = \frac{m}{2} \int_0^1 (1-y)^{m-n} [(1+y)^n - (1-y)^n] \frac{1}{y} \, dy.$$

This is the representation we want.

We consider two special cases:

(a) $m = 2N$, $n = N$

Assume that there is an even number, $2N$, of balls in U_1 at the start. We want to know how long it takes, on the average, until there are N balls in each urn. We find

$$\mu_N = N \int_0^1 [(1-y^2)^N - (1-y)^{2N}] \frac{1}{y} \, dy.$$

It is found successively that

$$\mu_N = N \int_0^1 [1 - (1-y)^{2N}] \frac{1}{y} dy - N \int_0^1 [1 - (1-y^2)^N] \frac{1}{y} dy$$

$$= N \int_0^1 \sum_{k=0}^{2N-1} (1-y)^k \, dy - N \int_0^1 y \sum_{j=0}^{N-1} (1-y^2)^j \, dy$$

$$= N \sum_{k=0}^{2N-1} \frac{1}{k+1} - \frac{N}{2} \sum_{j=0}^{N-1} \frac{1}{j+1}.$$

Finally we obtain

$$\mu_N = N \sum_{j=1}^{N} \frac{1}{2j-1}.$$

If N is large, we have the approximation

$$\mu_N \approx N[\ln(2\sqrt{N}) + \tfrac{1}{2}\gamma],$$

where $\gamma = 0.5772\ldots$ is Euler's constant; see 'Symbols and formulas' at the beginning of the book.

(b) $n = m$

We want to know how long it takes, on the average, until all balls have
been transferred from U_1 to U_2. We obtain

$$\mu_m = \frac{m}{2} \int_0^1 [(1+y)^m - (1-y)^m] \frac{1}{y} \, dy .$$

It is seen that

$$\frac{\mu_m}{m} - \frac{\mu_{m-1}}{m-1} = \frac{2^{m-1}}{m}$$

and so

$$\mu_m = \frac{m}{2} \sum_{i=1}^{m} \frac{2^i}{i} .$$

Comparing (a) and (b) we arrive at the climax of this section: The
expected time required to attain the same number of balls in the two urns
is quite short compared to that required to have all balls transferred from
one urn to the other. For example, when $m = 10$, we have $\mu_5 \approx 8.9365$ and
$\mu_{10} \approx 1,186.5$ which is a fantastic difference. The approximation in (a) is
fairly good already for $N = 5$; it gives the value 8.9323.

We end the section by indicating a subject which may perhaps interest
some researchers. Analyse the *Pólya-Ehrenfest model* (baptized by us): At
the start, two urns contain 0 and m balls. At each time-point $t = 1, 2, \ldots$,
one of the balls in the urns is drawn at random, and is moved from the urn
it is taken from to the other urn, adding one more ball to the latter urn.
Hence the total number of balls increases from m to $m + 1$ to $m + 2$, etc.

References: Ehrenfest and Ehrenfest (1907), Kemeny and Snell (1960,
§7.3), Blom (1989a).

11.6   Ehrenfest's game

Because of its relationship to Ehrenfest's urn model we call the following
diversion *Ehrenfest's game*: Number $2N$ cards from 1 to $2N$. Two players
take N cards each. (Random distribution is not essential.) At each round,
one of the numbers 1 to $2N$ is chosen at random. The player who has the
card with that number gives this card to the other player. The game stops
when one of the players has no cards; he is the winner.

Using the results in the previous section, we shall prove that the mean
number g_N of rounds until the game ends is given by

$$g_N = \tfrac{1}{2}\mu_{2N} - \mu_N , \tag{1}$$

where
$$\mu_N = N \sum_{i=1}^{N} \frac{1}{2i-1}, \quad \mu_{2N} = N \sum_{i=1}^{2N} \frac{2^i}{i}.$$

For example, if $N = 3$, that is, if 6 cards are used, it is found that $g_N = 37$, which is a reasonable average number of rounds. It is advisable not to use more cards, for g_N increases very rapidly: For example, for N equal to 4, 5 and 10, g_N becomes approximately 149, 584 and 555,690, respectively!

We shall give a very compact proof of (1). (After all, the whole section is meant for fun and not for serious study.) As in the preceding section introduce the states $E_0, E_1, \ldots, E_{2N}$, where E_i is the state that player 1 has i cards. Also, let μ_n be the mean time required for the transition $E_0 \to E_n$.

Our problem is to determine the mean number g_N of rounds required for the transition $E_N \to$ first of E_0 and E_{2N}.

We determine g_N by an indirect method. Consider the transition $E_0 \to E_{2N}$. It can be divided into three stages as follows:

a. The transition $E_0 \to E_N$.
b. The transition $E_N \to$ first of E_0 and E_{2N}.
c. Another transition $E_0 \to E_{2N}$ with probability $\frac{1}{2}$.

The transition in (c) occurs if the transition in (b) goes to E_0. We are lucky enough to know the mean for all transitions except that in (b); see the preceding section. Taking expectations we obtain the relation

$$\mu_{2N} = \mu_N + g_N + \tfrac{1}{2}\mu_{2N}.$$

This leads to expression (1).

11.7   A pill problem

This is the first half of Problem E3429 (1991, p. 264 and 1992, p. 684) in *The American Mathematical Monthly*:

A certain pill bottle contains a large pills and b small pills initially, where each large pill is equivalent to two small ones. Each day the patient chooses a pill at random; if a large pill is selected, (s)he breaks the selected pill into two and eats one half, replacing the other half, which thenceforth is considered to be a small pill.

What is the expected number of small pills remaining in the bottle when the last large pill is selected?

An equivalent formulation of the pill problem is the following: An urn contains a white and b black balls. Draw one ball at a time at random from the urn. If a white ball is drawn, it is not returned; instead, a black ball is placed in the urn. If a black ball is drawn, it is thrown away. The drawings continue until there are no white balls left. Find the expectation of the number of black balls then remaining in the urn.

We shall give two solutions. In both cases we suppose that the balls are drawn at times $t = 1, 2, \ldots$. Let $T_{a-1}, T_{a-2}, \ldots, T_0$ be the time-points when the number of white balls has just decreased to $a - 1, a - 2, \ldots, 0$, respectively; set $T_a = 0$ as well.

(a) *First solution*

Let N_i be the number of black balls in the urn at time T_i and X_i the number of black balls drawn in the interval (T_i, T_{i-1}). We shall determine $E(N_i)$.

Taking $a = i$ and $b = N_i$ in Section 7.4, formula (4), we obtain

$$E(X_i | N_i) = \frac{N_i}{i+1},$$

and hence

$$E(X_i) = \frac{E(N_i)}{i+1}.$$

According to the rules of the drawing we have

$$N_i = N_{i+1} - X_{i+1} + 1,$$

and so, taking expectations and using the expression for $E(X_i)$ with i replaced by $i + 1$, we obtain the relation

$$E(N_i) = \frac{i+1}{i+2} E(N_{i+1}) + 1.$$

By an inductive argument it follows that

$$E(N_i) = (i+1) \left[\frac{b}{a+1} + \frac{1}{a} + \frac{1}{a-1} + \cdots + \frac{1}{i+1} \right],$$

where $i = 0, 1, \ldots, a - 1$. Letting $i = 0$ we obtain

$$E(N_0) = \frac{b}{a+1} + \frac{1}{a} + \frac{1}{a-1} + \cdots + 1,$$

which is the answer to the problem.

(b) *Second solution*

Consider the b original black balls and the a added black balls separately.

First, number all original black balls from 1 to b and let X be the number of original black balls still in the urn when the drawings stop at time T_0. Set

$$X = U_1 + U_2 + \cdots + U_b,$$

where U_i is 1 if the ith black ball is in the urn at T_0 and U_i is 0 otherwise. From the reasoning used in Subsection (a) of Section 7.4 we know that $U_i = 1$ with probability $1/(a+1)$. Hence the rv X has expectation $b/(a+1)$.

Second, consider the added balls. Let the ball added at time T_j receive number j, where $j = a-1, a-2, \ldots, 0$. Let Y be the number of these balls remaining in the urn when the drawing stops at T_0. We have

$$Y = V_{a-1} + V_{a-2} + \cdots + V_0,$$

where V_j is 1 if the added ball j is still in the urn at T_0, and 0 otherwise.

Just after time T_j there are j white balls in the urn. The probability that the ball with number j is drawn after these j white balls is $1/(j+1)$. Hence we obtain

$$E(V_j) = P(V_j = 1) = \frac{1}{j+1},$$

and it follows that Y has expectation

$$\frac{1}{a} + \frac{1}{a-1} + \cdots + 1.$$

Adding the expectations of X and Y we obtain the same answer,

$$\frac{b}{a+1} + \frac{1}{a} + \frac{1}{a-1} + \cdots + 1,$$

as in (a).

Interested readers may perhaps also study the distribution of the number of drawings performed until there are no white balls left.

The model can be modified in various ways, for example, in the following: An urn contains a white, b red and c black balls. If the ball drawn is white or red, it is not returned; instead, a black ball is placed in the urn. If the ball is black, it is thrown away. The drawings continue until one of the events 'no white balls left' and 'no red balls left' occurs. Find the expected number of black balls then remaining in the urn.

12
Cover times

'Cover times' is a comparatively new subject, which, we believe, has not yet entered into the textbooks. Using the terminology of graph theory, a cover time is the expected time required for visiting all vertices when a random walk is performed on a connected graph. Examples written in simple language are given in Section 12.1, and thereafter further examples follow. The chapter ends with some general results.

Two elementary articles on cover times have been written by Wilf (1989) and Blom and Sandell (1992).

12.1 Introduction

We use some terms taken from graph theory, which are later illustrated by examples. Let G be a graph with n vertices, numbered $1, 2, \ldots, n$, and e edges. The number of vertices and the number of edges are sometimes infinite. The graph is *connected*; that is, there are always paths, consisting of one or more edges, from any vertex to any other vertex. Vertex i is directly connected with d_i other vertices $(i = 1, 2, \ldots, n)$.

Consider the following random walk on G. At $t = 0$ a particle is at a certain *starting-vertex*, say at vertex i. At $t = 1$ the particle moves from vertex i with probability $1/d_i$ to one of the vertices with which this vertex is directly connected. This movement is repeated at $t = 2, 3, \ldots$.

Let m_{in} be the expected time until all vertices have been visited. We call m_{in} a *cover time*. Generally, we assume that the walk stops when all vertices have been visited. However, we sometimes let the walk continue until the particle returns to the starting-vertex. Let u_{in} be the expected time, counted from $t = 0$, until this happens. We call u_{in} an *extended cover time*. Clearly, we always have $u_{in} > m_{in}$.

It is sometimes of interest to consider the expected time m_{ik} until $k < n$ different non specified vertices have been visited, including the starting-vertex. We call m_{ik} a *partial cover time*. If after that we let the particle return to the starting-vertex, we obtain the *extended partial cover time* u_{ik}.

The different kinds of cover times defined above are sometimes independent of the starting-vertex. They can always be found by solving an appropriate system of linear equations, as explained in the following example.

Example 1. A special graph

Consider the graph in Figure 1. The particle starts at vertex 1, which is marked with a 'bullet'. We want to find the partial cover time m_{13}; that is, the expected time until $k = 3$ vertices have been visited, including the starting-vertex.

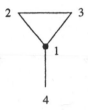

Fig. 1. A special graph.

When the particle leaves vertex 1, it moves to vertex 2, 3 or 4 with probability 1/3. Thus we have three cases:

1. The particle moves to vertex 2. Let a be the expected remaining time until one more vertex has been visited; then three vertices have been visited. The situation is illustrated in the leftmost graph in Figure 2; here a bullet indicates where the particle is at present, and a vertex already visited is encircled.
2. The particle moves to vertex 3. By symmetry, the expected remaining time is again a.
3. The particle moves to vertex 4. Let b be the expected remaining time.

Combining these three cases, we obtain the equation

$$m_{13} = 1 + \frac{2}{3} \cdot a + \frac{1}{3} \cdot b.$$

We then let the particle continue its walk from vertex 2, 3 or 4 until three vertices have been visited. Altogether we need four expected remaining times a, b, c, d corresponding to the graphs in Figure 2. This leads to the four additional equations

$$a = 1 + \frac{1}{2} \cdot 0 + \frac{1}{2} \cdot c,$$

$$b = 1 + d,$$

$$c = 1 + \frac{2}{3} \cdot 0 + \frac{1}{3} \cdot a,$$

$$d = 1 + \frac{2}{3} \cdot 0 + \frac{1}{3} \cdot b.$$

So we have five equations with five unknowns. We are only interested in m_{13} and find $m_{13} = 16/5$.

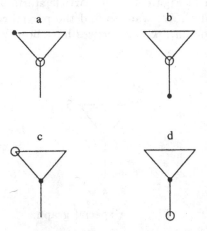

Fig. 2. Random walk on the special graph.

By a similar calculation, using 6 equations, we find the extended partial cover time $u_{13} = 76/15$. (We leave this calclation to the interested reader.)

As already said, the general method can be used in all cases. However, even for rather small values of n, the number of equations may be prohibitively large. In *The American Mathematical Monthly* Problem 6556 (1989, p. 847), the following question (closely related to cover times) is studied: Let a random walk take place on the eight corners of a cube. Find the mean number of steps required for visiting all *edges* at least once. This problem is solved by the general method, using a system with 387 unknowns! (The answer is ≈ 48.5.)

The general method is not always the only one available, as will be shown in several sections of this chapter.

There are many unsolved problems in this area. For example, let a particle wander on the integer-lattice points in the plane starting at the origin. At each moment $t = 1, 2, \ldots$ the particle moves to one of the four adjacent points (up, down, right or left) with the same probability $1/4$. It seems very difficult to find the partial cover time until k different points have been visited unless k is very small. It would be interesting to have an approximation valid for a large k. A related problem has been discussed by Aldous (1991).

Another unsolved problem is the following: Place a king somewhere on an empty chessboard. Let him move at random according to the usual rules until he has visited all 64 squares. Find the mean number of steps required. This problem seems extremely difficult. According to a simulation, consisting of one million rounds, the mean is approximately equal to

615. Here is another problem apt for simulation: does a knight cover the chessboard faster or slower than the king?

12.2   Complete graph

The graph with n vertices and $n(n-1)/2$ edges is called a *complete graph*; see Figure 1 for an example. A particle starts a random walk from one of the vertices and moves until all vertices have been visited. (See the preceding section for a complete description of the rules.) Find the cover time m_n, that is, the expected time until all vertices have been visited. Since the starting-vertex is unimportant here, we have dropped the index i in m_{in}.

It follows easily from Section 7.5 that the cover time is given by

$$m_n = (n-1)\Big(1 + \frac{1}{2} + \cdots + \frac{1}{n-1}\Big).$$

Fig. 1. Complete graph with four vertices.

We leave it to the reader to show that for $k \geq 2$ the partial cover time m_k and the extended partial cover time u_k are given by

$$m_k = (n-1)\Big(\frac{1}{n-k+1} + \frac{1}{n-k+2} + \cdots + \frac{1}{n-1}\Big),$$

$$u_k = m_k + (n-1).$$

12.3   Linear finite graph

A particle performs a symmetric random walk on the points $1, 2, \ldots, n$ of the x-axis. The points 1 and n are reflecting barriers; see Figure 1. We are interested in the cover time and the extended cover time.

(a) *Cover time*

First, suppose that the walk starts at the point 1. Apart from the terminology, we have considered this problem before. Formula (1), Section 10.2,
shows that

$$m_{1n} = (n-1)^2. \tag{1}$$

More generally, if i is the starting point we have

$$m_{in} = (i-1)(n-i) + (n-1)^2. \tag{2}$$

The proof is left to the reader.

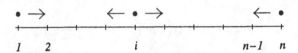

Fig. 1. Symmetric random walk on the line.

(b) *Extended cover time*

An excursion from point 1 to point n and back consists of two parts with
expected duration $(n-1)^2$, and so

$$u_{1n} = 2(n-1)^2. \tag{3}$$

More generally, it may be shown that

$$u_{in} = 2(n-i)^2 + 2(n-1)(i-1). \tag{4}$$

Again we leave the proof to the problem-minded reader.

12.4   Polygon

The reader of this section is assumed to be familiar with Section 10.9.

    Consider a polygon with n corners, numbered $0, 1, \ldots, n-1$, consecu-
tively. A particle starts from corner 0 and moves at $t = 1$ to one of the two
adjacent corners. The walk is continued in the same way at $t = 2, 3, \ldots$
until all n corners have been visited. The particle then continues the walk
until it returns to the starting point. We are interested in the cover time
and the extended cover time.

We begin with some general considerations. The walk on the polygon may be 'transformed' to a random walk on the x-axis starting from 0. We may thereafter apply the results of Section 10.9 concerning the range of this walk.

For example, if $n = 4$ the polygon is a square, and we then number the corners $-3, -2, -1, 0, 1, 2, 3$. Here corner -3 is the same as corner 1, -2 the same as 2, and so on. The walk on the x-axis continues until its range becomes equal to 4; the particle moving on the polygon has then visited all corners.

(a) *Cover time*

Let m_n be the cover time. (Since it is independent of where the walk starts, we need only one index.) Moreover, let T_n be the moment when the range becomes equal to n. By Section 10.9, Subsection (a), we obtain

$$m_n = E(T_n) = \frac{n(n-1)}{2}. \tag{1}$$

(b) *Extended cover time*

We express the extended cover time u_n as a sum

$$u_n = m_n + c_n, \tag{2}$$

where

$$c_n = \sum_{j=1}^{n-1} c_{nj} p_j.$$

Here p_j is the probability that corner j is the last corner visited and c_{nj} the expected time required for returning from this corner to corner 0.

The p's in this sum are all equal to $1/(n-1)$, which is indeed remarkable. (It might be expected that corners close to the starting point have a smaller chance of being the last corner visited, but this is not true!) To prove this statement, let the corresponding walk on the x-axis proceed until the range is R_{n-1}; then $n-1$ points have been visited. According to Section 10.9, Subsection (b), there are $n-1$ equiprobable alternatives for R_{n-1}. Hence the corner of the polygon not yet visited (that is, the last corner visited in the first part of the walk) is any of the $n-1$ corners $1, 2, \ldots, n-1$ with the same probability $1/(n-1)$.

Second, we consider the travel of the particle from corner j to corner 0, or the equivalent random walk on the x-axis. This can be seen as a classical random walk starting from a point j and stopping either at a point j steps to the right or $n - j$ steps to the left. The expected time until absorption is therefore given by $c_{nj} = j(n - j)$; see Section 1.5, Problem 2. The mean additional time until the particle returns to the starting point is therefore given by

$$c_n = \sum_{j=1}^{n-1} j(n-j) \cdot \frac{1}{n-1} = \frac{n(n+1)}{6}. \tag{3}$$

Inserting (1) and (3) into (2) we obtain the final answer

$$u_n = \frac{n(n-1)}{2} + \frac{n(n+1)}{6} = \frac{n(2n-1)}{3}.$$

In this chapter, we have only discussed symmetric random walks on graphs. Nonsymmetric random walks may also be of interest. We therefore invite the reader to study the walk of this section, assuming that the particle moves clockwise with probability p and counterclockwise with probability $q = 1 - p$.

12.5   A false conjecture

Consider a random walk on a connected graph with n vertices and e edges. Let the starting-vertex be given. How does the cover time m_n change if e varies? One might think that m_n is smaller when many vertices are connected by edges than when only a few vertices are connected, for in the former case it seems 'easier' for the particle to arrive at a new vertex.

For example, consider the two graphs in Figure 1. As usual, the starting point is indicated by a 'bullet'. [In (a) the choice of starting point is important, in (b) it is not.]

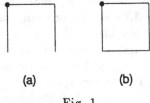

(a)   (b)

Fig. 1.

The graph in (a) has three edges and cover time $m_4 = 2 \cdot 1 + 3^2 = 11$, and the graph in (b) has four edges and cover time $m_4 = 4 \cdot 3/2 = 6$. (For the calculation of these quantities, see Sections 12.3 and 12.4, respectively.) So at least in this case the above rule applies. It may perhaps be tempting to believe that the rule is always true. Let us launch the following conjecture:

Conjecture: Draw an edge somewhere between two vertices not hitherto connected, and let m'_n be the cover time of the extended graph. Then $m'_n \leq m_n$.

The reader is invited to show that this conjecture is false by determining the cover time for the graph in Figure 2 and comparing it with the graph in Figure 1b.

Fig. 2.

However, the conjecture is not entirely wrong. The cover time has a tendency to decrease when e increases; see Figure 3.

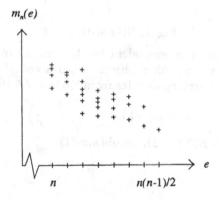

Fig. 3. Cover times $m_n(e)$ for graphs with n vertices and e edges.

12.6   Stars II

Stars and generalized stars were introduced in Section 10.5. A star is a graph with a central vertex surrounded by r vertices as shown in Figure 1. At $t = 0$ a particle is at the centre. At $t = 1$ it moves to one of the surrounding vertices chosen at random, and at $t = 2$ it returns to the centre. This movement is repeated at $t = 3, 4, \ldots$. We will show that the cover time m, that is, the expected time until all $r + 1$ vertices have been visited, is given by

$$m = 2r\left(1 + \frac{1}{2} + \cdots + \frac{1}{r}\right) - 1. \tag{1}$$

For this purpose, denote the time-points $0, 1, \ldots$ in the following way:

$$T_0, U_1, T_1, U_2, \ldots, T_{N-1}, U_N.$$

Here the U's are the time-points for the visits to a surrounding vertex and the T's the time-points for the visits to the central vertex. Note that $T_i = 2i$ and $U_i = 2i - 1$. The walk stops at U_N, at which point all vertices have been visited at least once. Here N is an rv, and m is the expected time from $T_0 = 0$ to U_N; that is, $m = E(U_N) = E(2N - 1)$.

Fig. 1. Star with $r = 4$.

We realize that N is equivalent to the number of drawings with replacement from an urn with r objects until a complete series has been obtained. As a conseqence, we infer from Section 7.5 that

$$E(N) = r\left(1 + \frac{1}{2} + \cdots + \frac{1}{r}\right).$$

Since we have $m = E(2N - 1)$, we obtain (1).

Fig. 2. Generalized star with $r = 4$, $n = 2$.

Here is a rather difficult problem: Consider a generalized star with r rays each having n vertices; see Figure 2. The end-vertices are reflecting. At $t = 0$ a particle is at the centre. At each time-point $t = 1, 2, \ldots$ the particle moves to one of the adjacent vertices with the same probability. Show that the cover time m is given by

$$m = 2r\left(1 + \frac{1}{2} + \cdots + \frac{1}{r}\right)n^2 - n^2.$$

12.7   Inequality for cover times

Let G be a connected graph with n vertices and e edges. Start the walk at vertex i. Let m_{in} be the cover time and u_{in} the extended cover time. The following inequality holds:

$$m_{in} < u_{in} \leq 2e(n-1). \tag{1}$$

As seen from formula (3) in Section 12.3, the constant 2 is the best possible. This result is due to Aleliunas et al. (1979); see also Palacios (1990).

The inequality (1) has an interesting consequence. Since, clearly, the number e of edges of G cannot have a larger order of magnitude than $n^2/2$, it follows from the inequality that the order of magnitude of m_{in} and u_{in} cannot be larger than n^3.

It is easy to check that this statement holds in the examples of graphs given earlier in this chapter. For example, the linear graph in Section 12.3 has cover time and extended cover time of order n^2.

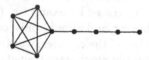

Fig. 1. Lollipop graph with 9 vertices.

The question arises whether graphs exist with cover times and extended cover times of order n^3. This is indeed the case, and as an example we take the following: Consider two graphs, a linear finite graph G_1 with n vertices and a complete graph G_2 with n vertices. Combine them into a single 'lollipop' graph G with $2n - 1$ vertices, as indicated in Figure 1. If the random walk starts at the vertex where the two graphs are coupled, the walk on G will have a cover time and an extended cover time of orders n^3 (this is also true for other starting points in G). The proof is not easy and will not be given here.

13
Markov chains

Markov chains affect the lives of students and professional probabilists in at least two ways.

First, Markov chains involve a special kind of dependence which is fruitful to study theoretically. Independence is a simple, clearly defined concept, but probabilistic dependence is a complex enterprise with many branch offices, Markovian dependence being one of the most important.

Second, Markov chains sometimes induce probabilists in a very healthy way to leave their ivory tower and participate in a multitude of practical applications of such chains in science, operations research, economics and elsewhere.

Father of the theory of Markov chains is the Russian mathematician *A.A. Markov* (1856–1922). It has later been extended to the more general field of Markov processes.

In our experience, many undergraduates acquire a rather scanty knowledge of Markov chains. Therefore, in Sections 13.1 and 13.2, certain basic results are reviewed and illustrated by examples. (Professional probabilists will find these sections boring and should only consult them for notation and terminology.) In the rest of the chapter, we give some more applications and study chains with doubly stochastic transition matrix and reversible chains, among other things.

There is an enormous amount of literature on Markov chains. Good sources are Kemeny and Snell (1960), Feller (1968), Isaacson and Madsen (1976) and Grimmett and Stirzaker (1992). Kemeny and Snell's book is a classic on finite chains, but its terminology differs from what is now standard.

13.1   Review I

We begin the review with an example that clarifies the nature of a Markov chain.

Example

In a country, there are two political parties, L and R. A person participates in elections at times $t = 0, 1, \ldots$. There are two *states*: $E_1 = $ 'vote for L' and $E_2 = $ 'vote for R'. If the person votes for L at time $t = n$, then with probability $1 - \beta$ he also votes for L at time $t = n + 1$. On the other hand, if he votes for R, then with probability $1 - \alpha$ he also votes for R at the next election. Earlier elections do not affect his behaviour. Under these conditions, the choice of party is governed by a Markov chain with *transition matrix*

$$P = \begin{pmatrix} 1 - \beta & \beta \\ \alpha & 1 - \alpha \end{pmatrix}.$$

The elements p_{ij} of P are called *1-step transition probabilities*. Here p_{ij} is the conditional probability that the person votes E_j given that he voted E_i in the previous election. Note that his choice is not affected by earlier elections; this is the fundamental Markov condition. It may be shown that the elements of the matrix product P^2 give the *2-step transition probabilities* $p_{ij}^{(2)} = P(E_i \rightarrow E_j$ in two steps). More generally, P^n gives the *n-step transition probabilities* $p_{ij}^{(n)}$.

Assume that, at his first election at $t = 0$, the ignorant citizen tosses a fair coin; then the *starting vector* of the chain is $p^{(0)} = (p_1(0), p_2(0)) = (\frac{1}{2}, \frac{1}{2})$. Also assume that in further elections he keeps to L with probability 0.9 and to R with probability 0.7. This means that $\alpha = 0.3$ and $\beta = 0.1$. At time $t = 1$ we obtain the *probability state vector* $p^{(1)} = p^{(0)}P$. Inserting the numerical values and performing the multiplication we find

$$(0.5 \quad 0.5) \begin{pmatrix} 0.9 & 0.1 \\ 0.3 & 0.7 \end{pmatrix} = (0.6 \quad 0.4).$$

Hence, at $t = 1$ the person votes L with probability 0.6 and R with probability 0.4. At time $t = 2$ the probability state vector is obtained from the relation

$$p^{(2)} = p^{(0)}P^2,$$

and analogously for $t = 3, 4, \ldots$.

The concepts introduced in this example are easily carried over to a general Markov chain with a finite number of states $E_1, \ldots, E_m$. We call this a *finite* Markov chain. (Chains with an infinite number of states also occur but are not discussed in this book.) The transition matrix P is then an $m \times m$ matrix $P = (p_{ij})$ with row sums 1. The probability state vector $p^{(n)}$ is a row vector with m elements obtained from the important relation

$$p^{(n)} = p^{(0)}P^n. \tag{1}$$

A chain is *irreducible* if it is possible to go from each state E_i to any other state E_j with positive probability, that is, if for all i and j the n-step transition probability $p_{ij}^{(n)}$ is positive for at least one value of n.

An irreducible chain is *periodic* with period d if n above is always a multiple of d. Here d is an integer at least equal to 2. Sections 13.2 and 13.3 contain examples of Markov chains with period $d = 2$. A chain which is not periodic is called *aperiodic*. An irreducible aperiodic finite Markov chain is called *ergodic*.

The two-state chain appearing in the election example is ergodic if $0 < \alpha < 1$ and $0 < \beta < 1$. (If $\alpha = \beta = 0$, the chain degenerates to an oscillating chain $E_1 E_2 E_1 E_2 \ldots$ with period 2.)

We end the section with a problem for the reader: Show that in the election example the matrix P^n has diagonal elements

$$p_{11}^{(n)} = \frac{\alpha}{\alpha + \beta} + \frac{\beta}{\alpha + \beta}(1 - \alpha - \beta)^n,$$

$$p_{22}^{(n)} = \frac{\beta}{\alpha + \beta} + \frac{\alpha}{\alpha + \beta}(1 - \alpha - \beta)^n.$$

13.2   Review II

(a) *Stationary distributions*

As stated in the previous section, the probability state vector at time $t = n$ can be found using the relation

$$p^{(n)} = p^{(0)} P^n. \tag{1}$$

Now assume that the probability row vector $\pi = (\pi_1, \ldots, \pi_m)$ is a solution of the equation

$$\pi = \pi P. \tag{2}$$

Using π as a starting vector, we find

$$p^{(1)} = p^{(0)} P = \pi P = \pi,$$

and by induction, generally for any n

$$p^{(n)} = \pi.$$

Hence the probability state vector is the same for all time-points n. The chain is then said to have a *stationary distribution* π.

A Markov chain can have several stationary distributions. However, if the chain is irreducible, the stationary distribution is unique.

(b) *Asymptotic distribution*

It is interesting to find out what happens to the chain when n tends to infinity. It may happen that there is a probability vector

$$\pi = (\pi_1, \ldots, \pi_m),$$

where all $\pi_j \geq 0$ and $\sum \pi_j = 1$, such that

$$p^{(n)} \to \pi$$

as $n \to \infty$, regardless of the starting vector $p^{(0)}$. We then say that the chain has an *asymptotic distribution* or *equilibrium distribution* π.

Using (1), it is not difficult to show that $p^{(n)}$ tends to π regardless of $p^{(0)}$ if and only if P^n tends to a matrix with all rows equal to the row vector π, that is, if

$$P^n \to \begin{pmatrix} \pi_1 & \pi_2 & \cdots & \pi_m \\ \pi_1 & \pi_2 & \cdots & \pi_m \\ \vdots & \vdots & \ddots & \vdots \\ \pi_1 & \pi_2 & \cdots & \pi_m \end{pmatrix}. \tag{3}$$

It can be shown that if an asymptotic distribution π exists, it is the solution of the system of $m + 1$ equations

$$\pi = \pi P; \qquad \sum_{j=1}^{m} \pi_j = 1. \tag{4}$$

This is the same system as that encountered in the case of stationary distributions. Hence, when the asymptotic distribution exists, it coincides with the stationary distribution, which is then unique.

(c) *Ergodic chain*

We already know from Section 13.1 that an irreducible aperiodic finite chain is called ergodic. If a Markov chain is ergodic, the asymptotic distribution π always exists. (The proof will not be given here.) In this case the probabilities π_i are always positive.

We are sometimes interested in the *recurrence time* N_{ii} for the first return to state E_i (by this we mean the number of steps required for the transition $E_i \to E_i$). If the chain is ergodic, the mean recurrence time is given by the simple expression

$$E(N_{ii}) = \frac{1}{\pi_i}. \tag{5}$$

(d) *Irreducible periodic chain*

When the chain is irreducible and periodic with period $d \geq 2$, the limiting property (3) does not hold. Instead we have

$$\frac{1}{d}\left(P^n + P^{n+1} + \cdots + P^{n+d-1}\right) \rightarrow \begin{pmatrix} \pi_1 & \pi_2 & \cdots & \pi_m \\ \pi_1 & \pi_2 & \cdots & \pi_m \\ \vdots & \vdots & \ddots & \vdots \\ \pi_1 & \pi_2 & \cdots & \pi_m \end{pmatrix} \tag{6}$$

where $\pi = (\pi_1, \pi_2, \ldots, \pi_m)$ is the solution of (4). Result (6) implies that

$$\frac{1}{d}\left[p^{(n)} + p^{(n+1)} + \cdots + p^{(n+d-1)}\right] \rightarrow \pi.$$

In words: the state probability vector $p^{(t)}$ averaged over $t = n, n+1, \ldots,$ $n + d - 1$ tends to π, for any choice of starting vector $p^{(0)}$. We then say that the chain has a *generalized asymptotic distribution*. Result (5) also holds for chains which are irreducible and periodic.

Example 1. *Ergodic chain*

Let us return to the election example in Section 13.1 with the transition matrix

$$P = \begin{pmatrix} 1 - \beta & \beta \\ \alpha & 1 - \alpha \end{pmatrix}$$

where $0 < \alpha < 1$ and $0 < \beta < 1$. The system of equations $\pi = \pi P$ becomes

$$\pi_1 = (1 - \beta)\pi_1 + \alpha\pi_2,$$
$$\pi_2 = \beta\pi_1 + (1 - \alpha)\pi_2.$$

We add the relation $\pi_1 + \pi_2 = 1$. The solution is

$$\pi_1 = \frac{\alpha}{\alpha + \beta}; \qquad \pi_2 = \frac{\beta}{\alpha + \beta}.$$

Here $\pi = (\pi_1, \pi_2)$ is both a stationary distribution and an asymptotic distribution.

Example 2. *Irreducible periodic chain*

Suppose that

$$P = \begin{pmatrix} 0 & 1 & 0 \\ q & 0 & p \\ 0 & 1 & 0 \end{pmatrix}.$$

The chain is irreducible and has period 2. Solving the system (4) we obtain $\pi = (q/2, 1/2, p/2)$. The probability state vector averaged over $t = n, t = n+1$ tends to π regardless of the starting vector.

Finally, we suggest a problem for the reader: Radio messages from Mars, studied by a Tellurian cryptanalyst, are written in the Martian language ABRACADABRA, which uses only one vowel A and four consonants BCDR. Interspersed in the ordinary communication, the cryptanalyst found some long messages generated by a Markov chain with the transition matrix given by

$$
\begin{array}{ccccc}
 & A & B & C & D & R \\
A & 0 & 1/2 & 1/2 & 0 & 0 \\
B & 0 & 1/3 & 0 & 1/3 & 1/3 \\
C & 0 & 0 & 1/3 & 1/3 & 1/3 \\
D & 1/2 & 0 & 0 & 0 & 1/2 \\
R & 1/2 & 0 & 0 & 0 & 1/2
\end{array}
$$

The cryptanalyst is interested in the distance X between consecutive vowels and in the distance Y between consecutive consonants in the Martian messages, assuming stationarity. Show that $E(X) = 9/2$ and $E(Y) = 9/7$.

13.3 Random walk: two reflecting barriers

A particle performs a random walk on the x-axis, starting from the origin and moving at $t = 1, 2, \ldots$, one step to the right or one step to the left among the points $x = 0, 1, \ldots, m$. If the particle is at $x = i$, it goes right with probability p_i and left with probability $q_i = 1 - p_i$. We suppose that $p_0 = 1$ and $p_m = 0$ and that the probabilities $p_1, \ldots, p_{m-1}$ are larger than 0 and smaller than 1. Hence the particle is reflected at the origin and at the point $x = m$.

The walk may be represented as a finite Markov chain with the states $E_0, \ldots, E_m$, where E_i corresponds to the point $x = i$. The transition matrix is an $(m + 1) \times (m + 1)$ matrix $P = (p_{ij})$, where $p_{i,i+1} = p_i$, $p_{i,i-1} = q_i$ and the remaining elements are zero. The chain is irreducible. Since the particle can only return to a given state after an even number of steps, the chain has period 2.

(a) *Stationary distribution*

When $m = 3$ we have

$$
P = \begin{pmatrix}
0 & 1 & 0 & 0 \\
q_1 & 0 & p_1 & 0 \\
0 & q_2 & 0 & p_2 \\
0 & 0 & 1 & 0
\end{pmatrix}.
$$

Solving the system of equations $\pi = \pi P$, we obtain the stationary distribution $\pi = (\pi_0, \ldots, \pi_3)$, where

$$\pi_k = \frac{p_0 p_1 \cdots p_{k-1}}{q_1 q_2 \cdots q_k} \pi_0 \tag{1}$$

for $k = 1, 2, 3$. The probability π_0 is obtained using the condition that the π's add up to 1. For a general m, the same expression holds for $k = 1, 2, \ldots, m$.

(b) *Generalized asymptotic distribution*

We apply what was said about generalized asymptotic distributions in Section 13.2. Since the chain has period 2, we have that, when $n \to \infty$,

$$\tfrac{1}{2}(P^n + P^{n+1}) \to \begin{pmatrix} \pi_1 & \pi_2 & \cdots & \pi_m \\ \pi_1 & \pi_2 & \cdots & \pi_m \\ \vdots & \vdots & \ddots & \vdots \\ \pi_1 & \pi_2 & \cdots & \pi_m \end{pmatrix}$$

where π is given in (a). Hence the probability state vector $p^{(t)}$ averaged over $t = n$ and $t = n + 1$ tends to π, and so π is a generalized asymptotic distribution.

We end the section with a problem for the reader: Assuming as before that the particle starts from the origin and performs a symmetric random walk among the points $x = 0, 1, \ldots, m$, show that the mean recurrence time until it returns to the origin is $2m$.

13.4   Ehrenfest's model II

Ehrenfest's urn model was introduced in Section 11.5. Let us repeat the basic facts. Two urns U_1 and U_2 together contain m balls. At $t = 1, 2, \ldots$ one of the m balls is drawn at random and is moved from the urn it is taken from to the other urn. Let E_k be the state that 'k balls are in urn U_2', where $k = 0, 1, \ldots, m$. The transition of the balls can be described by a Markov chain with these states.

The analysis of the chain is facilitated by the relationship with the random walk discussed in the preceding section. The state E_k is identified with 'particle at $x = k$'. In the present case we have $p_k = 1 - k/m$ and $q_k = k/m$. For example, if $m = 3$, the transition matrix is given by

$$P = \begin{pmatrix} 0 & 1 & 0 & 0 \\ 1/3 & 0 & 2/3 & 0 \\ 0 & 2/3 & 0 & 1/3 \\ 0 & 0 & 1 & 0 \end{pmatrix}.$$

(a) *Stationary distribution*

The stationary distribution $\pi = (\pi_0, \ldots, \pi_m)$ is obtained by inserting the values of p_k and q_k given above in the general expression (1)

$$\pi_k = \frac{p_0 p_1 \cdots p_{k-1}}{q_1 q_2 \cdots q_k} \pi_0$$

of the preceding section. After reduction we obtain

$$\pi_k = \binom{m}{k} \pi_0,$$

where $\pi_0 = (\frac{1}{2})^m$. Hence the final expression becomes

$$\pi_k = \binom{m}{k} \left(\frac{1}{2}\right)^m$$

for $k = 0, 1, \ldots, m$, which is a binomial distribution $\text{Bin}(m, \frac{1}{2})$. For example, if $m = 3$, we find $\pi_0 = 1/8$, $\pi_1 = 3/8$, $\pi_2 = 3/8$, $\pi_3 = 1/8$.

(b) *Generalized asymptotic distribution*

The Markov chain for the Ehrenfest model is irreducible and has period 2. The probability state vector $p^{(t)}$ averaged over $t = n$ and $t = n + 1$ tends to the binomial distribution $\text{Bin}(m, \frac{1}{2})$ regardless of the initial distribution of balls in the two urns. In terms of the original molecular model, each molecule will, in the long run, be in each receptacle with the same probability $\frac{1}{2}$, independently of the other molecules.

13.5 Doubly stochastic transition matrix

Assume that the transition matrix P of a Markov chain has the special property that not only the row sums but also the column sums are unity. The transition matrix is then said to be *doubly stochastic*.

An example is

$$P = \begin{pmatrix} \alpha & \beta & \gamma \\ \gamma & \alpha & \beta \\ \beta & \gamma & \alpha \end{pmatrix}.$$

If all elements are positive, the chain is ergodic. The asymptotic distribution $\pi = (\pi_1, \pi_2, \pi_3)$ in this example is obtained by solving the system of equations $\pi = \pi P$, $\sum \pi_i = 1$ or, explicitly,

$$\pi_1 = \alpha \pi_1 + \gamma \pi_2 + \beta \pi_3,$$
$$\pi_2 = \beta \pi_1 + \alpha \pi_2 + \gamma \pi_3,$$
$$\pi_3 = \gamma \pi_1 + \beta \pi_2 + \alpha \pi_3,$$
$$\pi_1 + \pi_2 + \pi_3 = 1.$$

The solution is $\pi = (1/3, 1/3, 1/3)$, and so the asymptotic distribution is uniform. This is true for any ergodic chain with doubly stochastic transition matrix.

We shall discuss two models for the transfer of balls between two urns, both involving a Markov chain with a doubly stochastic transition matrix. In order to simplify the discussion, the number of balls in the urns is small.

(a) *The Bernoulli–Laplace model*

There are four balls marked $1, 2, 3, 4$, two of which are placed initially in each urn. In each drawing, one ball is chosen at random from each urn and the balls change place.

Introduce a Markov chain with six states, denoted $1122, 1212, 1221,$ $2112, 2121, 2211$. The first number denotes the position of ball 1, the second number that of ball 2, and so on. (Hence 1122 means that balls 1 and 2 are in urn 1 and balls 3 and 4 in urn 2.) The transition matrix is found to be

$$P = \begin{pmatrix} 0 & 1/4 & 1/4 & 1/4 & 1/4 & 0 \\ 1/4 & 0 & 1/4 & 1/4 & 0 & 1/4 \\ 1/4 & 1/4 & 0 & 0 & 1/4 & 1/4 \\ 1/4 & 1/4 & 0 & 0 & 1/4 & 1/4 \\ 1/4 & 0 & 1/4 & 1/4 & 0 & 1/4 \\ 0 & 1/4 & 1/4 & 1/4 & 1/4 & 0 \end{pmatrix}.$$

The matrix is doubly stochastic, for the column sums are unity. The chain is ergodic and has a uniform asymptotic distribution over the six states. Hence in the long run, all states occur equally often.

(b) *Urns with indistinguishable balls*

There are now two indistinguishable balls in the urns. In each drawing, choose an urn at random and transfer a ball to the other urn. If the chosen urn is empty, do nothing.

Introduce a Markov chain with three states $0, 1, 2$ according to the number of balls in one of the urns. The transition matrix is

$$P = \begin{pmatrix} 1/2 & 1/2 & 0 \\ 1/2 & 0 & 1/2 \\ 0 & 1/2 & 1/2 \end{pmatrix}$$

and is doubly stochastic. The chain is ergodic and has a uniform asymptotic distribution over the three states.

As an exercise, we recast the welcoming problem in Section 1.6 into a Markov chain problem: A has a house with one front door and one back door. He places one pair of walking shoes at each door. For each walk, he chooses one door at random, puts on a pair of shoes, returns after the walk

to a randomly chosen door, and takes off the shoes at the door. If no shoes
are available at the chosen door, A walks barefoot. Show that the long run
probability that A performs a walk barefooted is $1/3$. (Hint: Introduce a
Markov chain with three states E_0, E_1, E_2, where $E_i = $ 'i pairs at the front
door before a walk begins'.)

Here is another problem. Consider a Markov chain with three states
E_1, E_2 and E_3 and transition matrix

$$\begin{pmatrix} 1/2 & 1/2 & 0 \\ 1/2 & 0 & 1/2 \\ 0 & 1/2 & 1/2 \end{pmatrix}.$$

Start the particle at E_1 and stop the process as soon as the particle returns
to E_1. Let Y_2 and Y_3 be the number of visits to E_2 and E_3 before the
stop. Prove that Y_2 and Y_3 have expectations 1. (The problem can be
generalized to a chain with n states: The mean number of visits to each of
$E_2, E_3, \ldots, E_n$ is 1; see Kemeny and Snell (1960, p. 121).)

13.6 Card shuffling

Methods of shuffling cards into random order have been discussed by many
probabilists, among them Poincaré (1912). Aldous and Diaconis (1986)
have investigated the following method:

Example. Top card in at random shuffle

Consider a deck of n cards in any initial order. Remove the top card and
insert it at random in one of the n possible positions. Remove the top card
again and insert it at random. Perform this procedure several times. The
deck will gradually tend to be in random order.

We now consider the following general procedure for shuffling n cards:
It is convenient to use Markov chain terminology. There are $m = n!$ states
$E_1, \ldots E_m$, each corresponding to a certain permutation of the n cards.
The cards are shuffled in several independent steps.

Suppose that the initial order, counted from top to bottom, is E_1. By
the first shuffling operation, E_1 is exchanged for some other state E_i, by
random choice. The next shuffling operation exchanges E_i for some other
state E_j, and so on. Thus the successive permutations obtained can be
described by a Markov chain with a certain $m \times m$ transition matrix P
(see the example below). If there are 52 cards in the deck, the transition
matrix is enormous for it has 52! states.

The transition matrix for this procedure is doubly stochastic (compare
the preceding section). This can be seen as follows: If at a certain stage

the chain shifts from E_j to E_k with probability p, then there is also a shift from some state E_i to E_j with the same probability. Since the elements in the jth row sum to unity, this must also be true for the elements in the jth column which proves the assertion.

Let us now assume that the shuffling procedure is such that the chain is ergodic. It then follows from Section 13.5 that the chain has a uniform asymptotic distribution over the $n!$ possible permutations. In other words: *In the limit, the cards will be ordered at random.* This is a marvellous conclusion!

We know that the chain is ergodic when it is aperiodic and irreducible; hence we must check any proposed shuffling procedure for these two properties.

Example. Top card in at random shuffle (continued)

Consider a deck with only three cards. There are then six states $E_1, \ldots, E_6$, which we call instead 123, 132, 213, 231, 312, 321, where we have ordered the cards from top to bottom. Suppose that the initial order is 123. After one step, the chain is in 123, 213 or 231, with the same probability $1/3$. Thus the first row of the transition matrix P is $1/3, 0, 1/3, 1/3, 0, 0$. The whole matrix is given by

$$
P = \begin{pmatrix}
1/3 & 0 & 1/3 & 1/3 & 0 & 0 \\
0 & 1/3 & 0 & 0 & 1/3 & 1/3 \\
1/3 & 1/3 & 1/3 & 0 & 0 & 0 \\
0 & 0 & 0 & 1/3 & 1/3 & 1/3 \\
1/3 & 1/3 & 0 & 0 & 1/3 & 0 \\
0 & 0 & 1/3 & 1/3 & 0 & 1/3
\end{pmatrix}.
$$

No periodicities occur, and each state can be attained sooner or later from each other state; hence the chain is ergodic. This implies that the shuffling method is satisfactory; in the long run it will lead to a random order of the cards.

An interesting question concerns the number of shufflings necessary until the order of the deck is sufficiently close to random. Much attention has been given to this challenging problem, which requires a lot of mathematics [see Bayer and Diaconis (1992) and the references quoted in that paper].

13.7 Transition times for Markov chains

Consider a Markov chain with transition matrix P. The transitions take place at $t = 1, 2, \ldots$. It is sometimes of interest to determine the *mean transition time* for the transition $E_i \rightarrow E_j$ where $i \neq j$. (Compare Section 13.2, where we mentioned the mean recurrence time for $E_i \rightarrow E_i$.)

More generally, we may divide the states of the chain into two groups of states G_1 and G_2 and determine the mean transition time from a given state E_i in G_2 to a nonspecified state in G_1; we call this the mean transition time from E_i to G_1. We present the solution of the latter more general problem without proof [see Kemeny and Snell (1960, Chapter III)].

Divide the transition matrix into four parts:

$$P = \begin{pmatrix} P_1 & S \\ R & Q \end{pmatrix}.$$

Here P_1 and Q show transition probabilities for the internal communication within G_1 and G_2, respectively, whereas R and S govern the transitions $G_2 \rightarrow G_1$ and $G_1 \rightarrow G_2$, respectively.

Let $E(N_i)$ denote the mean transition time from E_i in G_2 to G_1. To determine this mean we need the matrix $V = (v_{ij})$, where

$$V = (I - Q)^{-1}.$$

In V we retain the numbering of the rows and columns used in P. Here I is a unit matrix of the same size as Q. Then $E(N_i)$ is given by

$$E(N_i) = \sum v_{ij} \, ,$$

where the summation is performed for all j such that $E_j \in G_2$.

Example

A particle performs a random walk between the points E_1, E_2, E_3, E_4, E_5 given by, respectively,

$$(0,0), (0,1), (0,-1), (-1,0), (1,0).$$

The particle goes from the origin to one of the other four points with the same probability 1/4. From each of the other four points, the particle goes to one of the three adjacent points with the same probability 1/3. The walk stops as soon as the particle arrives at one of the points E_4 and E_5, whichever comes first. Find the mean transition time. See Figure 1.

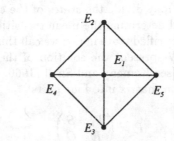

Fig. 1. A random walk in the plane.

The transition matrix P is given by

$$\begin{pmatrix} 0 & 1/4 & 1/4 & 1/4 & 1/4 \\ 1/3 & 0 & 0 & 1/3 & 1/3 \\ 1/3 & 0 & 0 & 1/3 & 1/3 \\ 1/3 & 1/3 & 1/3 & 0 & 0 \\ 1/3 & 1/3 & 1/3 & 0 & 0 \end{pmatrix}.$$

Take $G_1 = \{E_1, E_2, E_3\}$ and $G_2 = \{E_4, E_5\}$. We obtain

$$Q = \begin{pmatrix} 0 & 1/4 & 1/4 \\ 1/3 & 0 & 0 \\ 1/3 & 0 & 0 \end{pmatrix}; \quad I - Q = \begin{pmatrix} 1 & -1/4 & -1/4 \\ -1/3 & 1 & 0 \\ -1/3 & 0 & 1 \end{pmatrix}$$

and

$$(I - Q)^{-1} = \begin{pmatrix} 6/5 & 3/10 & 3/10 \\ 2/5 & 11/10 & 1/10 \\ 2/5 & 1/10 & 11/10 \end{pmatrix}.$$

The mean transition time from E_1 to the first of E_4, E_5 is obtained by adding the elements in the first row of this inverse:

$$\frac{6}{5} + \frac{3}{10} + \frac{3}{10} = \frac{9}{5}.$$

Alternatively, one may solve the problem by using a set of recursive relations.

Here is a similar problem for the interested reader: Let the transition matrix be

$$P = \begin{pmatrix} 1/3 & 1/3 & 1/3 \\ 1/4 & 1/2 & 1/4 \\ 1/4 & 1/4 & 1/2 \end{pmatrix}.$$

Show that the mean transition time from E_3 to E_1 is 4.

13.8 Reversible Markov chains

Consider an ergodic Markov chain with transition matrix $P = (p_{ij})$ and asymptotic distribution π. Hitherto, we have studied the future behaviour of the chain, given the past. We now change the perspective and look backwards from some time-point, say from $t = n$.

Let the chain have π as a starting vector so that $p^{(0)} = \pi$, which means that the chain is in equilibrium from the beginning.

First, assume that the chain is in state E_j at $t = n$. We want the conditional probability q_{jk} that it was in state E_k at $t = n - 1$. We find successively, using the rules of conditional probability,

$$q_{jk} = P(E_k|E_j) = \frac{P(E_k E_j)}{P(E_j)} = \frac{P(E_k)P(E_j|E_k)}{P(E_j)} = \frac{\pi_k p_{kj}}{\pi_j}.$$

(We have chosen this way of writing, in order to avoid cumbersome notation.)

Second, assume that the chain is in state E_i at $t = n$ and in state E_j at $t = n - 1$. We seek the probability that it was in E_k at $t = n - 2$. We find

$$P(E_k|E_j E_i) = \frac{P(E_k E_j E_i)}{P(E_j E_i)} = \frac{P(E_k)p_{kj}p_{ji}}{P(E_j)p_{ji}} = \frac{\pi_k p_{kj}}{\pi_j} = q_{jk}.$$

By an extension of this argument to more than three time-points, it is realized that, also in a backwards perspective, we obtain a Markov chain. We call this a *reversed Markov chain*; it has the transition matrix $Q = (q_{jk})$.

In the special case that $P \equiv Q$, the original chain is said to be a *reversible Markov chain*. This happens if and only if

$$\pi_j p_{jk} = \pi_k p_{kj} \tag{1}$$

for all j and k. In this case, the reversed chain is probabilistically indistinguishable from the original one.

We shall take an example which is familiar from Sections 13.1 and 13.2. Suppose that a Markov chain has the transition matrix

$$P = \begin{pmatrix} 1 - \beta & \beta \\ \alpha & 1 - \alpha \end{pmatrix}$$

where $0 < \alpha < 1$, $0 < \beta < 1$. We have shown before that the asymptotic distribution is given by $\pi = (\pi_1, \pi_2)$, where $\pi_1 = \alpha/(\alpha+\beta)$, $\pi_2 = \beta/(\alpha+\beta)$. We can see that

$$\pi_1 p_{12} = \frac{\alpha\beta}{\alpha + \beta}; \qquad \pi_2 p_{21} = \frac{\alpha\beta}{\alpha + \beta}.$$

Hence according to (1) the chain is reversible.

In order to find out whether a Markov chain is reversible we have to check condition (1) for all combinations of j and k. Alternatively, we may use *Kolmogorov's criterion*: A Markov chain is reversible if and only if, for any sequence of states, we have

$$
\begin{aligned}
P(E_{j_1} \rightarrow E_{j_2} \rightarrow \cdots \rightarrow E_{j_n} \rightarrow E_{j_1}) = \\
P(E_{j_1} \rightarrow E_{j_n} \rightarrow E_{j_{n-1}} \rightarrow \cdots \rightarrow E_{j_1}).
\end{aligned} \tag{2}
$$

A comprehensive discussion of reversible Markov chains is found in the book by Kelly (1979).

13.9   Markov chains with homesickness

Consider a Markov chain with states $E_1, E_2, \ldots, E_n$ and transition matrix $P = (p_{ij})$. If, for any $m \geq 2$ and any i, j,

$$
p_{ii}^{(2m)} \geq p_{ij}^{(2m)}, \tag{1}
$$

the chain is said to have *homesickness*. (For notation, see Section 13.1.)

This terminology, which we have chosen, may seem slightly frivolous but illustrates the behaviour of a particle starting from its home at E_i. It wants to return home as soon as possible and does so after $2m$ steps with a probability that is larger than, or possibly equal to, the probability of arriving at any other given state E_j after the same number of steps.

Mathematically speaking, homesickness is equivalent to the following: In any row of the product matrix P^{2m}, the diagonal element is larger than, or possibly equal to, any other element.

Example 1. Random walk on the x-axis

A particle performs a random walk on the points $x = 1, 2, 3, 4$; see Figure 1. If the particle is at $x = 2$ or $x = 3$, it moves one step to the left or one

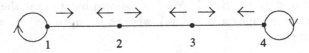

Fig. 1. Random walk on the x-axis.

step to the right with the same probability $\frac{1}{2}$. If it is at $x = 1$, it stays
there or goes one step to the right with the same probability $\frac{1}{2}$. Similarly,
if the particle is at $x = 4$, it stays there or goes one step to the left with the
same probability $\frac{1}{2}$. The transition matrix of the Markov chain generated
in this way is given by

$$P = \begin{pmatrix} 1/2 & 1/2 & 0 & 0 \\ 1/2 & 0 & 1/2 & 0 \\ 0 & 1/2 & 0 & 1/2 \\ 0 & 0 & 1/2 & 1/2 \end{pmatrix}.$$

This chain has homesickness as will be proved later. For the moment, we
only consider P^4:

$$P^4 = \begin{pmatrix} 3/8 & 1/4 & 1/4 & 1/8 \\ 1/4 & 3/8 & 1/8 & 1/4 \\ 1/4 & 1/8 & 3/8 & 1/4 \\ 1/8 & 1/4 & 1/4 & 3/8 \end{pmatrix}.$$

In each row, the diagonal element is larger than the other three elements.

We now leave this example and consider a general class of chains with
homesickness. The following result is due to Alon and Spencer (1992,
p. 139):

A Markov chain has homesickness, that is, satisfies (1), if

1. the transition matrix P is symmetric,
2. each row can be obtained from any other row by permutation of the
   elements.

Clearly, because of the symmetry, each column can be obtained from
any other column by permutation as well. In addition, note that the transition matrix is doubly stochastic, compare Section 13.5.

For the proof of Alon and Spencer's result we need the following neat
inequality probably due to Chebyshov: If $a_1, a_2, \ldots, a_n$ are given real numbers and $\pi = (\pi_1, \ldots, \pi_n)$ is a permutation of $1, 2, \ldots, n$, then

$$\sum_{i=1}^{n} a_i a_{\pi_i} \leq \sum_{i=1}^{n} a_i^2. \tag{2}$$

The inequality follows by expanding the left side of the inequality

$$\sum_{i=1}^{n} (a_i - a_{\pi_i})^2 \geq 0.$$

Since P is symmetric, P^r is symmetric as well. Using this fact and inequality (2), we obtain successively

$$p_{ik}^{(2m)} = \sum_{j=1}^{n} p_{ij}^{(m)} p_{jk}^{(m)} = \sum_{j=1}^{n} p_{ij}^{(m)} p_{kj}^{(m)}$$

$$\leq \sum_{j=1}^{n} p_{ij}^{(m)} p_{ij}^{(m)} = \sum_{j=1}^{n} p_{ij}^{(m)} p_{ji}^{(m)}$$

$$= p_{ii}^{(2m)}.$$

Hence the chain has homesickness.

The homesickness expressed by (1) holds, of course, even when the number $2m$ of steps is large; however, it gradually dwindles away. This is realized as follows. Since the transition matrix of the chain is doubly stochastic, the asymptotic distribution over the n states is uniform. In other words, as time passes, the particle tends to dwell with the same probability $1/n$ in each state; hence it gradually forgets where it started, and, loosely speaking, the homesickness fades out.

A mathematical remark

It is interesting to observe that the product matrix P^4 in Example 1 has the same property as P itself: Its four rows contain the same elements in different orders. This is no coincidence, but a consequence of the following general statement:

Let C be the class of square matrices which are symmetric and whose rows contain the same elements in different orders. (Each element may be positive, negative or zero.) If $A \in C$, then, for any integer $k \geq 2$, $A^k \in C$.

The proof is left to readers who welcome a mathematical problem in this book, for a change. However, we want to put forth a warning. The problem is probably not very difficult for the full-fledged algebraist or 'combinatorist', but possibly a hard nut to crack for the average statistician or probabilist. Therefore, we give a hint which will serve as an excellent nutcracker: Represent the $n \times n$ matrix A in C as a linear combination

$$A = \sum_{i=1}^{n} a_i P_i .$$

Here the a's are the elements in the rows and the P's are symmetric permutation matrices. (A permutation matrix has a 1 in each row and in each column, and 0's in all other positions.)

14
Patterns

Patterns are attractive. An elementary introduction to this subject was given in Section 1.4. In this chapter, the level is somewhat more advanced, but still accessible for an undergraduate. The top intellectual level concerning patterns will be reached in Section 16.6, which is the most advanced section of this book.

A reference on a medium level is Blom and Thorburn (1982), where further articles are mentioned. Interesting examples involving patterns are found in some textbooks, for instance in Ross (1983, p. 28, 75, 96, 231).

14.1   Runs II

In Section 6.1 we discussed combinatorial runs. We shall now consider runs in sequences of random trials. Such runs were investigated long ago by de Moivre, Simpson and Laplace, among others.

An event A occurs with probability p in a single trial. Set $q = 1 - p$. Perform n independent trials. In this section we define a run of length r as a sequence of r A's obtained in immediate succession. In 1738, de Moivre posed the following problem: Let B_n be the event that at least one run of length r or more occurs in n trials. Find the probability of B_n.

This is no easy problem. The clever de Moivre solved it by deriving the generating function of the probabilities $P(B_n)$. We shall give another solution.

The event B_n occurs if and only if a run of the length r is completed at the nth trial or earlier. The earliest possible trial is, of course, the rth. Hence we have

$$P(B_n) = P(N \leq n) = \sum_{k=r}^{n} P(N = k),$$

where N is the number of trials until the first run of the exact length r occurs. If we can find the probability function of N, de Moivre's problem is solved.

Set

$$p_k = P(N = k)$$

for $k = r, r+1, \ldots$. We have the recurrence

$$p_1 = p_2 = \cdots = p_{r-1} = 0, \tag{1}$$

$$p_r = p^r, \tag{2}$$

$$p_k = qp_{k-1} + pqp_{k-2} + \cdots + p^{r-1}qp_{k-r}. \tag{3}$$

Relation (3) holds for $k = r+1, r+2, \ldots$. There is a close relationship between (1), (2), (3) and formulas (1), (2), (3), respectively, in Section 5.2, and the sets of relations may be proved in almost the same manner:

Formulas (1) and (2) are, of course, true. For relation (3) we use the same idea as in Subsection (c) of Section 5.2:

Let A^* be the complement of A. The event $N = k$ takes place if *either* the first trial results in A^* and the remaining $k-1$ trials in a run of length r at the end *or* if the first two trials result in AA^* and the remaining $k-2$ trials in a run of length r at the end *or* if the first three trials result in $AAA^*, \ldots$, *or* if the first r trials result in $AA \ldots AA^*$ and the remaining $k-r$ trials result in a run of length r at the end. Since $P(A) = p$, $P(A^*) = q$ and the trials are independent, this immediately leads to (3).

We shall now use (1), (2), (3) for deriving the probability generating function $G(s) = E(s^N)$ of N.

We start from

$$p_r = p^r,$$

$$p_{r+1} = qp_r,$$

$$p_{r+2} = qp_{r+1} + pqp_r,$$

$$\vdots$$

$$p_{2r} = qp_{2r-1} + pqp_{2r-2} + \cdots + p^{r-1}qp_r,$$

$$p_{2r+1} = qp_{2r} + pqp_{2r-1} + \cdots + p^{r-1}qp_{r+1},$$

$$\vdots$$

Multiplying the rows by $s^r, s^{r+1}, \ldots$, respectively, and adding, we obtain the relation

$$G(s) = p^r s^r + qsG(s) + pqs^2 G(s) + \cdots + p^{r-1}qs^r G(s).$$

Solving for $G(s)$, performing a summation and a rearrangement, we obtain

$$G(s) = \frac{p^r s^r (1 - ps)}{1 - s + qp^r s^{r+1}}. \tag{4}$$

This formula is due to Laplace; it is closely related to that obtained for $P(B_n)$ by de Moivre. These results of our 'probabilistic ancestors' are impressive.

Let us finish the section with a problem for the reader: Using (4), or a direct approach, show that the expectation of the waiting time N is given by

$$E(N) = \frac{1}{p} + \frac{1}{p^2} + \cdots + \frac{1}{p^r}.$$

References: Feller (1968, p. 322), Hald (1990, p. 417).

14.2 Patterns II

We shall present some general results without proof concerning patterns in random sequences.

Draw random digits from the set $\{1, 2, ..., m\}$, one at a time. Continue until the last k digits form a given pattern S. We are interested in the *waiting time* N until S is obtained for the first time, and particularly in the *mean waiting time* $E(N)$.

We need some general concepts describing the *overlaps* that may occur within the pattern. For example, if $S = (1\ 2\ 1\ 2)$, there is an overlap 1 2 of length 2. In $S = (1\ 2\ 1\ 2\ 1)$, there are two overlaps, 1 and 1 2 1, of lengths 1 and 3, respectively. If the last r digits of S equal the r first, we say that there is an overlap of length r. To describe the overlaps we use the *leading numbers*

$$\epsilon_r = \begin{cases} 1 & \text{if there is an overlap of length } r \\ 0 & \text{if there is no overlap of length } r \end{cases}$$

for $r = 1, 2, \ldots, k$. Note that we always have $\epsilon_k = 1$. For example, the pattern $S = (1\ 2\ 1\ 2\ 1)$ has the leading numbers $\epsilon_1 = 1$, $\epsilon_2 = 0$, $\epsilon_3 = 1$, $\epsilon_4 = 0$, $\epsilon_5 = 1$.

The following quantity can be seen as a measure of the amount of overlap:

$$e = \sum_{r=1}^{k} \epsilon_r m^r. \tag{1}$$

We introduce the function

$$d(x) = \sum_{r=1}^{k} \epsilon_r x^r. \tag{2}$$

The probability generating function $G_N(s) = E(s^N)$ of the waiting time N is given by

$$G_N(s) = \frac{1}{1 + (1 - s)d(m/s)}. \tag{3}$$

This elegant formula is valid for all patterns. We may obtain the expectation of N by differentiating with respect to s and setting $s = 1$ in the derivative; compare Section 3.2. After some computation we find the simple expression

$$E(N) = d(m). \qquad (4)$$

An alternative form is

$$E(N) = e, \qquad (5)$$

where e is the amount of overlap.

For example, if the digits are binary and $S = (1\ 1)$, then $m = 2$ and $\epsilon_1 = \epsilon_2 = 1$. From (4) we obtain $E(N) = 2 + 2^2 = 6$, as found in Section 1.4 by another method.

It follows from (5) that $E(N)$ is smallest when the pattern is non-overlapping; then $e = m^k$ and $E(N) = m^k$.

Let us apply what we have found to letters chosen at random, which is a situation sometimes perpetrated by philosophically minded authors. An ape is given a typewriter with one key for each of the 26 letters A,B,C,...,Z of the alphabet. If the ape presses the keys at random, it will sooner or later hit on Shakespeare's collected works, but the mean waiting time, until this marvellous sequence is obtained, is enormous.

It is tempting to believe that the mean waiting time is the same for any given pattern of k letters. However, this is not true. Let us demonstrate this fact in some detail, limiting our attention to words. It will take the ape 26^3 letters, on the average, to get APE, but slightly more, $26 + 26^3$, to get DAD. It will take him $26 + 26^4 + 26^7$ letters to obtain ENTENTE or to obtain ALFALFA. It will take him 26^{11} letters, on the average, to get SHAKESPEARE, but $26 + 26^4 + 26^{11}$ to get ABRACADABRA.

There is a large literature on patterns. Some early results were obtained by Condorcet (see Todhunter (1865, p. 363)) and by Whitworth (1901). Two more recent papers are Li (1980) and Blom and Thorburn (1982). We shall describe Li's results in Section 16.6.

We end the section with a problem for the reader: Using (3), show that the variance of the waiting time is given by

$$Var(N) = [d(m)]^2 + d(m) - 2md'(m),$$

where $d'(x)$ denotes the first derivative of $d(x)$.

14.3   Patterns III

Draw random digits from the set $\{1, 2, \ldots, m\}$ until the last k digits form one of a given set of different patterns $S_1, \ldots, S_n$ of the same length k.

We are interested in the *mean waiting time* $E(N)$ until this happens and in the *stopping probabilities* $\pi_1, \ldots, \pi_n$ that the procedure ends in $S_1, \ldots, S_n$, respectively, where $\sum \pi_i = 1$.

We shall first consider the overlaps within and between the patterns. If the last r digits of S_i are equal to the first r digits of S_j, we say that there is an overlap of length r between S_i and S_j (in this order). In order to describe this overlap we use the leading numbers

$$\epsilon_r(i,j) = \begin{cases} 1 & \text{if there is an overlap of length } r \\ 0 & \text{if there is no overlap of length } r. \end{cases}$$

Here i, j assume the values $1, 2, \ldots, n$ and $r = 1, \ldots, k$. Note that, for all i and j, each ϵ_k is 1 if $i = j$ and 0 if $i \neq j$.

Two patterns may be more or less overlapping. The following quantity can be seen as a measure of the amount of overlap between S_i and S_j (in this order):

$$e_{ij} = \sum_{r=1}^{k} \epsilon_r(i,j) m^r; \tag{1}$$

compare (1) in the preceding section, which is a special case obtained for $i = j$.

Example 1

If the digits are binary and $S_1 = (1\,1\,1\,1)$, $S_2 = (2\,1\,1\,1)$, we have:

r	$\epsilon_r(1,1)$	$\epsilon_r(1,2)$	$\epsilon_r(2,1)$	$\epsilon_r(2,2)$
1	1	0	1	0
2	1	0	1	0
3	1	0	1	0
4	1	0	0	1

A simple calculation shows that $e_{11} = 30$, $e_{12} = 0$, $e_{21} = 14$, $e_{22} = 16$.

Returning to the general case, the mean waiting time $E(N)$ and the stopping probabilities π_j are obtained by solving the system of equations

$$\sum_{j=1}^{n} e_{ji}\pi_j = E(N), \qquad i = 1, 2, \ldots, n,$$

$$\sum_{j=1}^{n} \pi_i = 1. \tag{2}$$

A proof of this result (written in a different form) is given by Blom and Thorburn (1982). Also see Section 16.6.

Special case

When there are only two patterns S_1 and S_2, the linear system (2) reduces to

$$e_{11}\pi_1 + e_{21}\pi_2 = E(N),$$
$$e_{12}\pi_1 + e_{22}\pi_2 = E(N), \tag{3}$$
$$\pi_1 + \pi_2 = 1.$$

Hence

$$\frac{\pi_1}{\pi_2} = \frac{e_{22} - e_{21}}{e_{11} - e_{12}}, \tag{4}$$

which shows the odds for obtaining S_1 first. Inserting the π's in the first or the second equation we obtain $E(N)$.

Example 1 (continued)

Let us compute the probabilities π_1 and π_2 that we get S_1, or S_2, first. It follows from (4) that

$$\frac{\pi_1}{\pi_2} = \frac{16 - 14}{30 - 0} = \frac{1}{15}.$$

Hence $\pi_1 = 1/16, \pi_2 = 15/16$. Furthermore, from the first or second equation (3) we obtain that $E(N) = 15$.

   Here is a problem for the interested reader: Throw a symmetrical die until a multiple run of length 3 is obtained, that is, one of the patterns $S_1 = (1\,1\,1), S_2 = (2\,2\,2), \ldots, S_6 = (6\,6\,6)$. First show that $e_{ij} = 6 + 6^2 + 6^3 = 258$ when $i = j$, and $e_{ij} = 0$ otherwise. Use this result to show that the mean waiting time is 43.

   It is, of course, possible to study patterns with different lengths. Consider, for example, the following problem: Toss a symmetric coin until $S_1 = (1\,0)$ or $S_2 = (0\,1\,1)$ occurs. With the same notation as above, show that $\pi_1 = 3/4, \pi_2 = 1/4, E(N) = 7/2$.

14.4   A game for pirates

A man was taken prisoner by pirates, who did not know what to do with him. Finally the captain decided to write the letters L I V E K on a die, leaving one side blank. The die was to be thrown until one of the words LIVE or KILL was formed by four consecutive letters. The pirates, who liked to gamble, were enthusiastic about the idea. The captain asked if the prisoner had any last wish before the gambling started. 'Yes', he said, 'I would be glad if you could replace the word KILL by DEAD'. The captain agreed and wrote the letters L I V E A D on the die.

In this way the clever prisoner increased his chance of survival from $124/249 \approx 0.4980$ to $217/433 \approx 0.5012$. However, this small change of the gambling rule increases the mean waiting time for a decision from $78,125/249 \approx 313.8$ to $281,232/433 \approx 649.5$. We leave the proof to the reader; use the results of the previous section and be careful and patient!

14.5   Penney's game

This section demands some acquaintance with game theory; see for example Owen (1968).

Penney (1969) has proposed the following game: Players A_1 and A_2 toss a fair coin. Denote heads by 1 and tails by 0. Let k be a given positive integer. Before the game, A_1 chooses a sequence S_1 of k 1's and 0's. He shows it to A_2, who is invited to choose another sequence S_2 of the same length. The coin is tossed until S_1 or S_2 has appeared; in the first case, A_1 is the winner; in the second case, A_2 wins.

We are interested in the optimal sequences for A_1 and A_2, that is, their best choices of patterns. In order to determine these sequences, we must be able to calculate the winning probabilities of the players for any given pair of sequences.

(a) *Winning probabilities*

For given S_1 and S_2, let π_1 and π_2 be the winning probabilities of A_1 and A_2. We apply the results in Section 14.3 with $m = 2$ and calculate $e_{11}, e_{12}, e_{21}, e_{22}$ according to (1) of that section. From (4) in the same section we obtain

$$\frac{\pi_1}{\pi_2} = \frac{e_{22} - e_{21}}{e_{11} - e_{12}}, \qquad (1)$$

which shows A_1's odds for winning the game.

Example

Let us take $S_1 = (1\ 0)$ and $S_2 = (0\ 0)$. We have

$$e_{11} = 4; \quad e_{12} = 2; \quad e_{21} = 0; \quad e_{22} = 6,$$

and from (1) we obtain A_1's odds $\pi_1 : \pi_2 = 3 : 1$. Hence $\pi_1 = 3/4$ and $\pi_2 = 1/4$.

(b) *Some game theory*

For choosing the best patterns we need a result from game theory. Denote the $n = 2^k$ possible sequences of length k by $S_1, S_2, \ldots, S_n$.

Let p_{ij} be the probability that A_1 wins the game if he chooses S_i and A_2 chooses S_j. According to the fundamental theorem for finite zero-sum games, the best sequences are those for which the max min of the p's, that is,

$$\max_i \min_j p_{ij}$$

is attained.

(c) *Application: $k = 2$*

For sequences of length 2 there are four possibilities

$$S_1 = (1\,1); \quad S_2 = (1\,0); \quad S_3 = (0\,1); \quad S_4 = (0\,0).$$

Because of symmetry it suffices to let A_1 choose between the first two of these. Each of these patterns is combined with the other three, and the winning probability of A_1 is calculated for each pair. The result is:

	S_1	S_2	S_3	S_4
S_1	–	1/2	1/4	1/2
S_2	1/2	–	1/2	3/4

In the first row the smallest value is 1/4, and in the second it is 1/2. Since the largest of these values is 1/2, this is the max min; the corresponding optimal sequences for A_1 and A_2 are either (1 0), (0 1) or (1 0), (1 1); the game is then fair. (By symmetry, the two other optimal sequences are (0 1), (1 0) and (0 1), (0 0).)

(d) *Application: $k \geq 3$*

When the patterns have lengths ≥ 3, the game is always unfavourable for A_1; for a proof see Chen and Zame (1979). Below we give below optimal sequences when $k = 3$ and 4; the solution is, in general, not unique.

$k = 3$: A_1 and A_2 should take (1 0 0) and (1 1 0), respectively; A_1 wins with probability $1/3 \approx 0.3333$.

$k = 4$: Optimal sequences are (1 0 1 1) and (0 1 0 1); A_1 wins with probability $5/14 \approx 0.3571$.

A reader endowed with energy and patience is invited to prove the result given above for $k = 3$; compare Martin Gardner's column in *Scientific American*, October 1974, where help can be obtained.

14.6   Waldegrave's problem II

In Section 5.2 we considered Waldegrave's problem: $r + 1$ persons play a sequence of rounds in the 'circular' order

$$A_1, \ldots, A_{r+1}, A_1, \ldots, A_{r+1}, \ldots$$

until one player has won over all the others in immediate succession. Each round is won with probability $\frac{1}{2}$.

We shall now determine the expectation of the number N of rounds until the game stops. For this purpose, we may use formula (3) in Section 5.2, but it is more convenient to apply the theory of patterns. We then use the same trick as in Subsection (c) of Section 5.2. Let us quote from that section:

Associate with each round, from the second and onwards, a 1 if the winner of the round also won the preceding round, and a 0 otherwise. The game stops when $r - 1$ 1's in succession have been obtained, that is, at the first occurrence of a run of length $r - 1$.

We now apply the results obtained in Section 14.2, using formulas (2) and (4). Taking $k = r - 1$ and $\epsilon_1 = \epsilon_2 = \cdots = \epsilon_k = 1$, formula (2) shows that

$$d(x) = x + x^2 + \cdots + x^{r-1} = \frac{x^r - x}{x - 1}.$$

Thus, we obtain from (4) that

$$E(N') = d(2) = 2^r - 2,$$

where N' is the number of rounds counted from the second. Adding the first round, we obtain

$$E(N) = 2^r - 1.$$

Thus we have learned that the theory of runs in random sequences can advantageously be seen as a part of the more general theory of patterns.

14.7   How many patterns?

A sequence of n numbers is collected at random from the set $\{1, 2, \ldots, m\}$. Hence each number from the set appears with probability $p = 1/m$ at a given position in the sequence. Let S be a given pattern of length $k \geq 2$. Count the number X of S's that appear in the sequence. We are interested in the properties of the rv X.

The smallest possible value of X is 0 and the largest is $n - k + 1$. For example, suppose that we perform eight tosses with a balanced coin; we then have $m = 2$ and $n = 8$. Call heads 1 and tails 2 and suppose that we are interested in the pattern $S = (1\ 1\ 1)$. If all eight tosses result in heads, 1 1 1 1 1 1 1 1, we have $X = 6$; this is the largest possible value in this example.

We shall determine the expectation and variance of X, and also have a look at its distribution. These three tasks increase in difficulty: the expectation is simple to find, the variance is more complicated, and to derive an explicit expression for the exact distribution is generally a formidable enterprise.

(a) *Expectation and variance*

Set

$$X = U_1 + U_2 + \cdots + U_{n-k+1},\tag{1}$$

where U_i is 1 if S occurs at positions $i, i+1, \ldots, i+k-1$ and U_i is 0 otherwise. Each U_i is 1 with probability p^k and 0 with probability $1 - p^k$, and hence has expectation p^k. The dependence properties of the U's are interesting: if U_i and U_j are at a distance such that $j - i \geq k$, they are independent.

In Section 9.2 we studied a model for an rv X with exactly these properties. So we may quote from (2) of that section that

$$E(X) = (n - k + 1)p^k.\tag{2}$$

Furthermore, if $n \geq 2k - 1$, we showed that

$$Var(X) = A + 2B,\tag{3}$$

where

$$A = (n - k + 1)Var(U_1),\tag{4}$$

$$B = (n - k)Cov(U_1, U_2) + (n - k - 1)Cov(U_1, U_3) + \cdots$$
$$+ (n - 2k + 2)Cov(U_1, U_k).\tag{5}$$

We shall determine the variances and covariances appearing in (4) and (5). Since U_1^2 and U_1 have the same expectation p^k, we find

$$Var(U_1) = E(U_1^2) - [E(U_1)]^2 = p^k - p^{2k}.$$

This is the same result as in Section 9.2. However, the covariances are more complicated. We find

$$Cov(U_1, U_{j+1}) = E(U_1 U_{j+1}) - E(U_1)E(U_{j+1})$$
$$= E(U_1 U_{j+1}) - p^{2k},\tag{6}$$

where $1 \leq j \leq k - 1$. The product $U_1 U_{j+1}$ is 0 unless both U_1 and U_{j+1} are 1, which happens if S occurs at positions $1, 2, \ldots, k$ and at positions $j+1, j+2, \ldots, j+k$. This is possible only when the last $k - j$ digits in S are the same as the first $k - j$ digits, that is, when S has an overlap of this length. Using the ϵ-notation introduced in Section 14.2, this is equivalent to stating that $\epsilon_{k-j} = 1$. Given such an overlap, the rv's U_1 and U_{j+1} are both 1 with probability p^{k+j}. If S has no overlap of this length, the product $U_1 U_{j+1}$ is zero. Thus we have proved that

$$E(U_1 U_{j+1}) = \epsilon_{k-j} p^{k+j}.$$

From (6) we obtain the covariances, which are inserted in (5). The variance of X is given by (3) with

$$A = (n - k + 1)(p^k - p^{2k}), \tag{7}$$

$$B = \sum_{j=1}^{k-1}(n - k - j + 1)p^k(\epsilon_{k-j}p^j - p^k). \tag{8}$$

When S is nonoverlapping, the ϵ's in (8) vanish, which makes the variance simpler. In this special case, the variance is always smaller than the mean.

We shall give two very simple examples. Perform $n = 3$ tosses with a fair coin. First, consider the nonoverlapping pattern $S = (1\ 2)$. We then have $E(X) = 1/2$, $A = 3/8$, $B = -1/16$ and $Var(X) = 1/4$. Second, consider the overlapping pattern $S = (1\ 1)$. We find $E(X) = 1/2$, $A = 3/8$, $B = 1/16$ and $Var(X) = 1/2$.

(b) Distribution

As said before, it is generally complicated to derive the exact distribution of the number X of patterns. However, approximations can sometimes be used.

Set

$$\lambda = (n - k + 1)p^k. \tag{9}$$

The relations (2), (3), (7), (8) derived above show that

$$E(X) = \lambda, \tag{10}$$

$$Var(X) = \lambda - \lambda p^k + 2\lambda \sum_{j=1}^{k-1}\left(1 - \frac{j}{n - k + 1}\right)(\epsilon_{k-j}p^j - p^k). \tag{11}$$

Let c be the quotient of $Var(X)$ and $E(X)$.

Suppose that n is large compared to k and that p^k is small so that the probability $P(U_i = 1)$ is small and the mean $E(X)$ is of moderate size. There are two cases:

Case 1. $c \approx 1$

This case is encountered when the pattern is nonoverlapping or when the overlapping is small. The rv X is then approximately Poisson distributed. For nonoverlapping patterns one can prove that the following simple bound holds for the variation distance between X and an rv Y having a Poisson distribution with mean λ:

$$d_V(X, Y) \leq (2k - 1)p^k.$$

(The variation distance d_V was defined in Section 8.3.)

Case 2. $c > 1$

This case occurs when the overlapping is not small. Clusters of overlapping copies of the pattern then occur with rather high probability, and a Poisson approximation is not very good; it can be shown that approximation with a compound Poisson distribution may be much better.

For a more complete discussion on these matters and proofs of the above statements see Barbour, Holst and Janson (1992, p. 162).

Example 1

Collect $n = 1000$ random letters from the set of 26 letters $\{A, B, \ldots, Z\}$ and consider the nonoverlapping pattern $S = (A\,B)$. The quantity $p^k = (1/26)^2 = 1/676$ is small, $E(X) = 999/676 \approx 1.4778$, $Var(X) \approx 1.4734$ and $c \approx 0.9970$ is near 1. Therefore, a Poisson distribution with mean $999/676$ is a reasonable approximation of the distribution of X.

Example 2

Perform $n = 1000$ tosses with a fair coin and count the number X' of clusters of at least nine consecutive heads. That number is almost the same as the number X of the nonoverlapping pattern 'tail followed by nine heads'; X and X' are equal unless the sequence of tosses starts with nine heads, which happens with the small probability $(\frac{1}{2})^9 = 1/512$. As in Example 1 the rv X is approximately Poisson distributed, now with mean $E(X) = 991(\frac{1}{2})^{10}$. This approximation can also be used for X', but it is better to use a Poisson distribution with mean

$$E(X') = (\tfrac{1}{2})^9 + E(X) = 993(\tfrac{1}{2})^{10} \approx 0.9697,$$

thus taking into account the possibility that the sequence starts with nine heads.

Example 3

Perform the coin-tossing as in Example 2 and count the number Y of runs of nine consecutive heads. We have $E(Y) = 992(\frac{1}{2})^9 \approx 1.9375$. As this pattern has a 'maximal' overlapping, a Poisson approximation of Y is not good; there is clustering of overlapping patterns. A compound Poisson distribution taking account of the clustering is a better approximation.

We can argue as follows: After each pattern of nine consecutive heads, the number of tosses Z until the first tails has a geometric probability function $(\frac{1}{2})^k$ for $k = 1, 2 \ldots$. Therefore, the number of the pattern 'nine consecutive heads' in each cluster has approximately this distribution with mean 2 ('approximately' because only a fixed number of tosses are made). The number of clusters is the rv X' studied in Example 2. Making an adjustment so that the approximation has the same mean as Y, we can conclude: the distribution of Y might be approximated by the compound Poisson distribution given by the rv

$$Y' = Z_1 + Z_2 + \cdots + Z_N,$$

where $N, Z_1, Z_2, \ldots$ are independent, N is Poisson with mean $992(\frac{1}{2})^{10}$ and the Z's are geometrically distributed rv's, all with mean 2.

Here is an exercise for the reader: The probability that more than 1000 tosses with a fair coin are required to obtain nine consecutive heads is approximately 0.38. Show this.

15
Embedding procedures

'Embed' is a magic word in probability theory which opens a door between continuous and discrete probability. One may sometimes tackle a hard problem in continuous probability by embedding a discrete one in it, which is easier to solve. The opposite can also occur—a discrete problem may be solved by embedding it in a continuous setting. In this chapter we deal only with the second alternative; we shall give alternative solutions to several of the problems considered in earlier chapters.

The price the reader must pay for being introduced to this area of probability theory is that (s)he must be acquainted with order statistics from the exponential and gamma distributions. David (1981) and Arnold, Balakrishnan and Nagaraja (1992) are good references. An article on various embedding procedures is by Blom and Holst (1991).

15.1   Drawings with replacement

We shall use embedding in the following situation: An urn contains m balls, m_i of which have colour i, where $i = 1, 2, \ldots, s$. Draw balls one by one at random with replacement. Other drawing procedures are possible; see Section 15.5 for drawings without replacement. Use the following stopping-rule: Let $q_1, q_2, \ldots, q_s$ be non-negative integers. When q_i balls of colour i have been obtained, this colour is said to have obtained its *quota*. The drawings stop when k nonspecified colours have obtained their quotas. We are interested in the number N of drawings made, and especially in the mean and variance of N.

This problem is quite general and includes many questions concerning fair coins, symmetric dice, decks of cards and random numbers.

The drawings are embedded in s independent Poisson processes in the following way: If the first drawing results in a ball of the ith colour, we associate this drawing with the first jump of the ith process. If the second drawing results in a ball of another colour, say the jth, we associate this drawing with the first jump of the jth process, and so on. (Details are given later.) We assume that the ith process has intensity m_i/m and that it performs its jumps at the time-points $X_{i1} < X_{i2} < \cdots$; see Figure 1. Let

$$Y_{i1} = X_{i1}; \quad Y_{i2} = X_{i2} - X_{i1}; \quad \ldots$$

be the time intervals between the jumps; it follows from the properties of the Poisson process that the Y's are independent with $Y_{ij} \sim \text{Exp}(m/m_i)$.

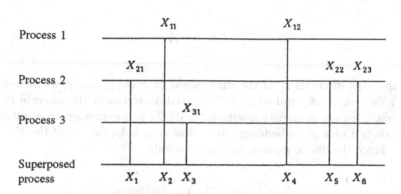

Fig. 1. Construction of superposed Poisson process.

Let T_i be the time of the q_ith jump of the ith process; at this moment the ith colour reaches its quota. Evidently, we have

$$T_i = \sum_{j=1}^{q_i} Y_{ij}.$$

(When $q_i = 0$ we set $T_i = 0$.) From a well known property of addition of independent exponential rv's with the same mean, we have

$$T_i \sim \Gamma(q_i, m/m_i).$$

The stopping time T when k of the processes have reached their quotas, and the drawings stop, is the kth order statistic of the rv's $T_1, \ldots, T_s$, that is, $T = T_{(k)}$, where

$$T_{(1)} < T_{(2)} < \cdots < T_{(s)}.$$

Hence the distribution of T is known in principle from the theory of order statistics for the gamma distribution. Note that T is a function of *independent* rv's T_i, which is one of the highlights of the embedding method.

We now construct another representation of T: Let us superpose the s independent Poisson processes and thus construct a single Poisson process; see Figure 1. This process has an intensity which is the sum of the intensities of the individual processes; hence it has intensity 1. It is obvious how balls are associated with the jumps of the superposed process. In Figure 1, X_1 corresponds to a ball of colour 2, X_2 to a ball of colour 1, and so on.

Intuitively, it appears that the sequence of colours generated by the superposed process is probabilistically equivalent to the sequence generated by the drawings with replacement from the urn. A strict proof of this can be found, for example, in Blom and Holst (1991).

Consequently, the stopping time T can be represented as a sum

$$T = \sum_{v=1}^{N} Y_v$$

of the time intervals Y_v of the superposed process. (Clearly, we have $Y_1 = X_1$, $Y_2 = X_2 - X_1$ and so on.) This relation connects the discrete rv N, in which we are primarily interested, with T, and expresses, in a nutshell, the central idea of embedding. Note that N is independent of the Y's.

From the above representation we obtain

$$E(T) = E\left(\sum_{v=1}^{N} Y_v\right) = E(N)E(Y_1).$$

Since $Y_1 \sim \text{Exp}(1)$ we have $E(Y_1) = 1$ and the basic relation

$$E(N) = E(T). \tag{1}$$

Higher moments can also be obtained. Since the moment generating function of the Y's is $1/(1-t)$, we obtain

$$E(e^{tT}) = E\left[(1-t)^{-N}\right].$$

Developing both sides in series we find

$$\sum_{j=0}^{\infty} \frac{t^j}{j!} E(T^j) = \sum_{j=0}^{\infty} \frac{t^j}{j!} E[N(N+1)\cdots(N+j-1)].$$

(The expectations on the right are the ascending factorial moments of N.) From this relation it follows, for example, that

$$E[N(N+1)] = E(T^2). \tag{2}$$

Hence, using (1) and (2), the variance of N can be expressed in terms of the mean and variance of the continuous rv T.

15.2 Repetition of colours

We now give an application of the embedding method described in the previous section.

An urn contains m balls, each ball having its special colour. At each drawing, one ball is selected at random and is then returned to the urn. The drawings continue until the first repetition of any one colour occurs, that is, until one of the balls has been obtained twice. We are interested in the number N of balls drawn until this happens.

In the general urn scheme in Section 15.1 we take

$$s = m, \ m_1 = m_2 = \cdots = m_s = 1, \ q_1 = q_2 = \cdots = q_m = 2, \ k = 1.$$

The stopping time of the Poisson processes is given by

$$T = \min(T_1, T_2, \ldots, T_m),$$

where $T_i \sim \Gamma(2, m)$. We now use a relation between the gamma and the Poisson distribution: If $Y \sim \text{Po}(x/m)$, we have

$$P(T_i > x) = P(Y \leq 1),$$

or, explicitly,

$$\frac{1}{m^2} \int_x^\infty y e^{-y/m} dy = \frac{x}{m} \cdot e^{-x/m} + e^{-x/m}.$$

(Use partial integration for the proof.) From this relation we obtain successively

$$P(T_i > x) = \left(1 + \frac{x}{m}\right) e^{-x/m},$$

$$P(T > x) = \left(1 + \frac{x}{m}\right)^m e^{-x},$$

$$E(T) = \int_0^\infty \left(1 + \frac{x}{m}\right)^m e^{-x} dx.$$

When m is large, we have, approximately,

$$\left(1 + \frac{x}{m}\right)^m \approx \exp\left(x - \frac{x^2}{2m}\right),$$

using the first two terms in the Taylor expansion of $\ln(1 + x/m)$. After a small reduction we obtain

$$E(N) = E(T) \approx \int_0^\infty e^{-x^2/2m} dx = \sqrt{\frac{\pi m}{2}}. \tag{1}$$

Furthermore, we find (compare the exercise at the end of Section 3.1)

$$E[N(N+1)] = E(T^2)$$

$$= \int_0^\infty 2x\left(1+\frac{x}{m}\right)^m e^{-x}dx \tag{2}$$

$$\approx \int_0^\infty 2xe^{-x^2/2m}dx = 2m.$$

The approximations can be refined by adding one more term when approximating $(1+x/m)^m$. Leaving out the calculations, we obtain

$$E(N) \approx \sqrt{\frac{\pi m}{2}} + \frac{2}{3}, \tag{3}$$

$$E[N(N+1)] \approx 2m + \sqrt{2\pi m}. \tag{4}$$

From relations (3) and (4) we may obtain the variance of N.

15.3   Birthdays revisited

We shall apply the findings in the preceding section to the birthday problem already studied in Sections 7.1, 8.1 and 9.1. Suppose that each year has 365 days and that births are distributed at random throughout the year. Select one person at a time until r people are found with the same birthday. Let N be the number of people selected.

(a) *Two people with the same birthday*

Let us take $r = 2$. Using the exact procedure described in Problem 2 in Section 7.1, it is found that the mean number $E(N)$ of people selected is 24.617 (to three decimal places). Taking $m = 365$ in the approximation formula (3) in Section 15.2, we obtain 24.611. The standard deviation is 12.192, and approximations (3) and (4) in the same section give 12.147.

(b) *Three people with the same birthday*

Using the same method as in Section 15.2 we obtain [since $T_i \sim \Gamma(3, m)$]

$$E(N) = \int_0^\infty \left[1 + \frac{x}{m} + \frac{1}{2}\left(\frac{x}{m}\right)^2\right]^m e^{-x}dx.$$

The integrand is approximately equal to $\exp[-x^3/(6m^2)]$. (We have again used a Taylor expansion.) Integration shows that

$$E(N) \approx 6^{1/3}\Gamma\left(\frac{4}{3}\right)m^{2/3} \approx 1.623m^{2/3}.$$

Taking $m = 365$ we find $E(N) \approx 82.875$. The exact value (to three decimal places) is 88.739.

15.4 Coupon collecting III

In a food package there is a prize belonging to one of two sets A_1 and A_2. Each set consists of n different types. A person buys one package at a time until either set A_1 or set A_2 is complete. Find the expected value of the number of packages which he has bought.

In Section 7.6 we gave a solution of this problem. We shall now give another solution, using the embedding method.

There are in this case $2n$ separate Poisson processes, each with intensity $1/2n$. Corresponding to each set, let $T_{11}, \ldots, T_{1n}$ and $T_{21}, \ldots, T_{2n}$ be the time-points of the first jump, that is, T_{ij} is the time-point when, for the first time, the person obtains a prize of type j in set A_i. Note that the T's are independent and $\text{Exp}(2n)$. The stopping time can be written as

$$T = \min(X_{1n}, X_{2n}),$$

where

$$X_{in} = \max(T_{i1}, \ldots, T_{in})$$

for $i = 1, 2$. We can now proceed in two different ways:

(a) *First derivation*

The rv X_{in} has the distribution function $(1 - e^{-x/2n})^n$ and so the rv T has the distribution function

$$P(T \leq x) = 1 - [1 - (1 - e^{-x/2n})^n]^2.$$

Hence the expectation of T can be written

$$E(T) = \int_0^\infty [1 - (1 - e^{-x/2n})^n]^2 dx = 2nJ_n,$$

where

$$J_n = \int_0^\infty [1 - (1 - e^{-x})^n]^2 \, dx.$$

Introducing the substitution $u = 1 - e^{-x}$, we obtain

$$J_n = \int_0^1 \frac{(1 - u^n)^2}{1 - u} \, du$$

$$= \int_0^1 (1 + u + \cdots + u^{n-1})(1 - u^n) \, du = \cdots$$

$$= \left(1 + \frac{1}{2} + \cdots + \frac{1}{n}\right) - \left(\frac{1}{n+1} + \frac{1}{n+2} + \cdots + \frac{1}{2n}\right).$$

Finally, this leads to the answer

$$E(N) = E(T) = 2n\left(1 + \frac{1}{2} + \cdots \frac{1}{n}\right) - 2n\left(\frac{1}{n+1} + \cdots + \frac{1}{2n}\right).$$

For a large n, $E(N)$ is approximately equal to $2n[\ln(n/2) + \gamma]$, where $\gamma = 0.5772\ldots$ is Euler's constant.

(b) *Second derivation*

Here we use a trick; we represent T as

$$T = X_{1n} + X_{2n} - \max(X_{1n}, X_{2n}).$$

Hence we find after some consideration (which is a challenge to the reader!)

$$E(T) = 2n\left(2\sum_{j=1}^{n}\frac{1}{j} - \sum_{j=1}^{2n}\frac{1}{j}\right).$$

This leads to the same answer as before.

The reader may perhaps want to solve the corresponding problem for three sets, each consisting of n different types of prizes.

Reference: Blom and Holst (1989).

15.5   Drawings without replacement

Consider the same urn scheme as in Section 15.1 with the modification that drawings are now performed without replacement. As before, we draw balls one at a time until some stopping-rule applies. We want to find the mean and variance of the number, N, of balls drawn. We will show that the embedding method can be used even in this case; however, we can no longer use Poisson processes for the embedding.

The reader is assumed to be familiar with the beta distribution and with order statistics and spacings from a continuous uniform distribution $U(0,1)$.

Embed the drawings in the order statistics from $U(0,1)$ in the following way: Imagine that the m drawings are made at time-points $X_1 < X_2 < \cdots < X_m$ corresponding to an ordered random sample of m values from $U(0,1)$. Associate with each such time-point the colour of the ball drawn at that moment. The ordered sample then splits into s independent ordered samples

$$X_{i1} < X_{i2} < \cdots < X_{im_i},$$

one for each colour. The ith colour reaches its quota at the moment X_{iq_i}. The procedure stops at the moment T when the kth order statistic of the times X_{iq_i}, $i = 1, \ldots, s$, occurs. Thus the distribution of T is known, in principle.

On the other hand, we can represent T as a sum

$$T = \sum_{v=1}^{N} Y_v$$

of the N spacings $Y_1 = X_1, Y_2 = X_2 - X_1, \ldots, Y_N = X_N - X_{N-1}$.

Since the spacings have mean $1/(m+1)$, we obtain the basic relation

$$E(N) = (m+1)E(T) \tag{1}$$

between the expectations of N and T.

The second and higher moments can also be obtained. For given $N = n$, the rv T has a beta distribution with density function proportional to $x^{n-1}(1-x)^{m-n}$ (see 'Symbols and formulas' at the beginning of the book), and so

$$E(T^2 | N = n) = \frac{n(n+1)}{(m+1)(m+2)}.$$

Taking the expectation with respect to N we find

$$E[N(N+1)] = (m+1)(m+2)E(T^2). \tag{2}$$

Hence $E(N)$ and $Var(N)$ can be derived from relations (1) and (2).

An application is given in the next section.

15.6   Socks in the laundry

In *The American Mathematical Monthly*, Problem E3265 (1988, p. 456) the following problem is posed:

Suppose that n pairs of socks are put into the laundry. Each sock has precisely one mate. After the load is done, socks are drawn out at random one at a time.

Suppose that the first pair is realized in draw number N. Find $E(N)$.

Applying the method of the preceding section, we take $m = 2n$, $s = n$, $m_1 = \cdots = m_n = 2$, $q_1 = \cdots = q_n = 2$ and $k = 1$. We find

$$T = \min(X_{12}, X_{22}, \ldots, X_{n2}),$$

where each X_{i2} is the largest of two independent values from U(0,1). Hence each such rv has the distribution function x^2 and so (since the X's are independent)

$$P(T > t) = P(\text{all } X_{i2} > t) = (1 - t^2)^n.$$

Therefore,

$$E(T) = \int_0^1 (1 - t^2)^n \, dt,$$

and so by (1) in the preceding section

$$E(N) = (2n + 1) \int_0^1 (1 - t^2)^n \, dt.$$

This is a well-known integral, and the final answer becomes

$$E(N) = \frac{(2n)(2n - 2) \cdots 2}{(2n - 1)(2n - 3) \cdots 1} = \frac{(2^n n!)^2}{(2n)!}.$$

According to Stirling's formula, we have for large n that $E(N) \approx \sqrt{\pi n}$.

When $n = 100$ the mean is 17.74 and the approximation gives 17.72. When $n = 1000$ (which is, of course, a very large number in this problem), the approximation yields 56.05. Thus, only about 56 drawings are required, on the average, to find a mate.

Here is a related problem: solve the Martian version. A Martian has three feet, and so three socks are needed for a complete set.

15.7   Negative hypergeometric II

An urn contains $m = a + b$ balls, a white and b black. Balls are drawn at random one at a time without replacement until the first white ball is obtained. It was proved in Section 7.4 that the number N of drawings required has expectation

$$E(N) = \frac{a + b + 1}{a + 1}.$$

We now give an alternative proof using the elegant embedding technique. We start from relation (1) in Section 15.5:

$$E(N) = (a + b + 1)E(T),$$

where T is the time-point when the last colour reaches its quota. Denoting white and black by 1 and 2, and taking $q_1 = 1, q_2 = 0, k = 2$, we have

$$T = \max(T_1, T_2),$$

where T_i is the time-point when colour i reaches its quota, $i = 1, 2$. Since $q_2 = 0$, we have $T_2 = 0$ and hence

$$T = T_1.$$

As the drawings are embedded in an ordered sample from U(0,1), and $q_1 = 1$, it follows that T_1 is the smallest of a independent U(0,1) variables. Then T_1 has distribution function $1 - (1 - x)^a$ and so $E(T_1) = 1/(a + 1)$. It follows from the basic relation that

$$E(N) = \frac{a + b + 1}{a + 1}.$$

We end this section with a challenge for the reader: Let N_k be the number of drawings necessary to obtain the kth white ball. Determine $E(N_k)$.

15.8   The first-to-r game I

Two persons A and B participate in the following game: At the start, A has r white cards and B has r black ones. At the first round, a card is selected at random among the $2r$ cards and is removed. At the second round, a card is selected among the $2r - 1$ remaining cards and is removed, and so on. The game continues until one of the players has no cards left; he is the winner.

It is instructive to illustrate the game by a random walk in the first quadrant; compare Section 10.1. A particle starts from the origin and goes a step to the right if a white card is drawn and a step upwards if a black card is drawn. The walk continues until either the line $x = r$ or the line $y = r$ is attained; in the former case A wins, in the second case B wins.

Let N be the number of rounds. We shall determine $E(N)$ by three different methods.

(a) *First method*

We proved in Subsection (c) of Section 7.4 that N has probability function

$$P(N = k) = 2\binom{k-1}{r-1} \Big/ \binom{2r}{r}$$

for $k = r, r+1, \ldots, 2r-1$. We can see that

$$E(N) = \sum_{k=r}^{2r-1} k \, 2 \binom{k-1}{r-1} \Big/ \binom{2r}{r} = 2r \sum_{k=r}^{2r-1} \binom{k}{r} \Big/ \binom{2r}{r}$$

$$= 2r \left[\sum_{k=r}^{2r} \binom{k}{r} \Big/ \binom{2r}{r} - 1 \right].$$

Noting that the terms $P(N = k)$ sum to unity, we obtain the identity

$$\sum_{k=r}^{2r-1} \binom{k-1}{r-1} = \frac{1}{2} \binom{2r}{r}.$$

Replacing r here with $r+1$, we obtain

$$\frac{1}{2} \binom{2r+2}{r+1} = \sum_{k=r+1}^{2r+1} \binom{k-1}{r} = \sum_{k=r}^{2r} \binom{k}{r}.$$

Inserting this sum in the expression for $E(N)$ we obtain, after a reduction, the simple expression

$$E(N) = \frac{2r^2}{r+1}.$$

(b) *Second method*

Consider the moment when the game stops. Suppose that A at that point has X cards left and B has Y cards left. If A wins, then $X = 0$, and if B wins, then $Y = 0$. Evidently, we may represent N as

$$N = 2r - X - Y. \tag{1}$$

For example, take $r = 4$ and suppose that the cards have been drawn in the order (with easily understood symbols)

$$B \; A \; B \; A \; B \; B \; (A \; A),$$

where the two last A's are the remaining cards. In this case B wins and $N = 6$, $X = 2$, $Y = 0$.

We need the expectations of X and Y. In order to find $E(X)$, write a permutation of all $2r$ cards as

$$\ldots B \ldots B \ldots B \ldots B \ldots$$

where we have only marked the B's explicitly. The r A's are divided by the B's into $r+1$ groups of equal mean size $r/(r+1)$. (This can be proved in a strict manner using the theory of exchangeable events; compare also Section 7.4.) Since $E(X)$ is the mean size of the last group, we have

$$E(X) = \frac{r}{r+1},\tag{2}$$

and similarly for $E(Y)$. Inserting this in (1), we obtain

$$E(N) = 2r - \frac{2r}{r+1} = \frac{2r^2}{r+1}.$$

Another way of proving (2) is the following: Number the r white cards from 1 to r. Introduce zero–one rv's $U_1, U_2, \ldots, U_r$ such that U_k is 1 if the kth white card is drawn after all the r black cards, and U_k is 0 otherwise. Evidently,

$$P(U_k = 1) = \frac{1}{r+1}$$

for $k = 1, 2, \ldots, r$.

Hence we have $E(U_k) = 1/(r+1)$. As

$$X = U_1 + \cdots + U_r,$$

we obtain (2).

(c) *Third method*

We use the embedding procedure described in Section 15.5. In the present problem we have $m = 2r$, $s = 2$, $m_1 = m_2 = r$, $q_1 = q_2 = r$ and $k = 1$.

The ordered random sample of $2r$ values from the uniform distribution $U(0,1)$ splits into two independent ordered samples

$$X_{i1} < X_{i2} < \cdots < X_{ir}$$

for $i = 1, 2$. The sampling procedure stops at the moment

$$T = \min(X_{1r}, X_{2r}).$$

Since X_{1r} and X_{2r} are the largest values in two samples of size r from $U(0,1)$ they have both distribution function x^r, hence

$$P(T > x) = (1 - x^r)^2.$$

Taking $m = 2r$ in the relation $E(N) = (m+1)E(T)$, we obtain

$$E(N) = (2r+1) \int_0^1 (1 - x^r)^2 \, dx,$$

and after a simple calculation

$$E(N) = \frac{2r^2}{r+1},$$

as obtained before.

In Section 17.1 a variant of the first-to-r game is studied.

We end the section with a problem for the reader: Three players have r cards each, and play according to similar rules as above until one of them has no cards left. Show that the expected number of rounds is given by

$$E(N) = \frac{6r^3(3r+1)}{(r+1)(2r+1)}.$$

16
Special topics

This chapter is, on the average, more difficult than the others. Sections 16.1–16.4 concern exchangeability, martingales and Wald's equation. These sections depend on deep results obtained by Olav Kallenberg, J. L. Doob and Abraham Wald, respectively, and the readers are invited to taste some appetizing fruits gathered by these outstanding researchers. Sections 16.1 and 16.4 contain two surprising results which should not be missed. Sections 16.5 and 16.6 are perhaps a bit long-winded and can be skipped, at least at the first reading.

16.1   Exchangeability III

Exchangeability for events was defined in Section 2.2 and will now be extended to rv's. We shall present an astonishing result concerning exchangeable rv's.

A finite sequence $X_1, X_2, \ldots, X_n$ or a denumerably infinite sequence $X_1, X_2, \ldots$ of rv's is said to be *exchangeable* if, for any k, the distribution of the k-dimensional rv

$$(X_{i_1}, X_{i_2}, \ldots, X_{i_k})$$

(where the subscripts are different) does not depend on the chosen subscripts, only on the dimension k.

Example 1. Drawings without replacement from an urn

An urn contains a white balls and b black balls. Balls are drawn at random without replacement until the urn is empty. Let $X_i = 1$ if a white ball is obtained at the ith drawing and $X_i = 0$ otherwise. The finite sequence $X_1, X_2, \ldots, X_{a+b}$ is exchangeable.

Example 2. Pólya's urn model.

An urn contains a white and b black balls. At each drawing, a ball is selected at random and returned together with c balls of the same colour. Define X_i as in Example 1. The infinite sequence $X_1, X_2, \ldots$ is exchangeable.

Exchangeable rv's are of great interest in probability theory since they are a natural generalization of independent rv's with a common distribution.

We need the concept of a stopping time. (It has been introduced earlier in this book, but we repeat the description.) Consider a sequence of rv's $Y_1, Y_2, \ldots$ which is finite or denumerably infinite. The rv N is a *stopping time* if for every k the event $N = k$ is determined by the outcome of $Y_1, Y_2, \ldots, Y_k$ and $\sum_k P(N = k) = 1$.

For example, perform independent tosses with a coin until the first heads is obtained. Suppose that this happens at the Nth toss; then N is a stopping time.

We shall consider sequences of exchangeable rv's with stopping times. A sequence of independent rv's has, of course, the same distribution following a stopping time as following a fixed index. This is also true for exchangeable sequences, which is indeed remarkable. This property was discovered by the Swedish probabilist Olav Kallenberg.

Example 1 (continued)

Draw balls from the urn until r white balls have been obtained, where $r < a$. Draw one more ball. Determine the probability that also this ball is white.

It is tempting to believe that, since the symmetry has been destroyed, the probability differs from the unconditional probability that a white ball is obtained in the first drawing. However, this is not so; the answer is $a/(a + b)$.

The reader is recommended to check this result by a direct calculation in a special case, say $a = 2$, $b = 2$, $r = 1$.

Example 2 (continued)

Draw balls according to the rule given until r white balls have been obtained. Draw one more ball. The probability is $a/(a + b)$ that it will be white.

The result in Example 1 above is a consequence of the following result: Consider a finite sequence $X_1, X_2, \ldots, X_n$ of exchangeable rv's. Let N be a stopping time defined on the sequence such that $P(N \leq n - k) = 1$. Then the k-dimensional rv's $(X_{N+1}, \ldots, X_{N+k})$ and $(X_1, \ldots, X_k)$ have the same distribution.

The result in Example 2 follows from a corresponding result for a general infinite sequence of exchangeable rv's: The rv's $(X_{N+1}, \ldots, X_{N+k})$ and $(X_1, \ldots, X_k)$ have the same distribution.

Simple proofs of the above statements have been given by Blom (1985). Kallenberg (1982) has discussed these questions in great generality. The

results obtained in the 1982 paper and elsewhere have the following astonishing consequence [quoted from Kallenberg (1985)]:

'Consider a lottery of the kind which can often be seen in the street or in
tivolies, where a fixed number of tickets are being sold, and the winners
receive their prizes immediately. Remaining tickets and prizes are both on
display, so that anyone can (at least in principle) keep track of the current
chances of winning or losing at every moment. *Suppose that a customer
has decided beforehand to buy a certain fixed number of tickets* (italicized
by the authors of the present book). What policy should he choose?

Intuitively, since the proportion of winning tickets will normally fluctuate a great deal, a good policy might be to wait until this proportion
exceeds a suitable level above the average. Of course there is always a
chance that this level will never be reached, and in that exceptional case
our customer could e.g. choose to buy the last few tickets from the lottery.
Other policies might be suggested, but it turns out that it does not make
any difference, since *the probability of getting a winning ticket will always
be the same.*'

The clue to this astonishing statement is found in the italicized sentence in
the first paragraph of the quotation.

The following elementary problem contains a special case of this general statement. In a lottery there are n tickets, two of which give prizes.
The tickets are sold one by one, and it is immediately announced if a buyer
wins. A and B buy tickets as follows: A buys the first ticket sold. B buys
the first ticket sold after the first winning ticket. Show that A's and B's
chances of winning a prize are the same.

16.2 Martingales

The theory of martingales is an important and fascinating part of probability theory, with many applications. The basic ideas are due to the
American probabilist Doob (1953). A good introduction to the subject is
given by Grimmett and Stirzaker (1992).

We begin with a new aspect of a familiar example.

Example 1. Gambling

A gambler participates in a game consisting of several rounds. In each
round he wins 1 or loses 1 with the same probability $\frac{1}{2}$. We associate with
the ith round an rv X_i which takes the values $+1$ and -1 depending on
whether the gambler wins or loses the round.

After n rounds, the player's profits are given by

$$Y_n = X_1 + X_2 + \cdots + X_n.$$

We observe that $Y_{n+1} = Y_n + X_{n+1}$. Hence, taking conditional expectations for given $X_1, \ldots, X_n$, we obtain

$$E(Y_{n+1}|X_1, \ldots, X_n) = E(Y_n|X_1, \ldots, X_n) + E(X_{n+1}|X_1, \ldots, X_n).$$

Since Y_n is determined by the rv's $X_1, \ldots, X_n$ and X_{n+1} is independent of these rv's, we find

$$E(Y_{n+1}|X_1, \ldots, X_n) = Y_n. \tag{1}$$

We now leave this example and turn to the general case. Relation (1) is then used as a definition of a martingale:

Let $X_1, X_2, \ldots$ and $Y_1, Y_2, \ldots$ be sequences of rv's. The sequence $\{Y_i\}$ is said to be a *martingale* with respect to the sequence $\{X_i\}$ if for all $n = 1, 2, \ldots$ we have

$$E(|Y_n|) < \infty$$

and

$$E(Y_{n+1}|X_1, \ldots, X_n) = Y_n.$$

Suppose now that we observe the X's until they satisfy some given stopping condition. Let N be the number of X's observed. We call N a *stopping time*. (We have encountered this concept several times earlier.)

For example, in the example with the gambler, he may stop playing when he has won one round. Then N is the smallest integer n such that $X_n = 1$.

Now suppose that

a. $P(N < \infty) = 1$,
b. $E(|Y_N|) < \infty$,
c. $E(Y_n|N > n)P(N > n) \to 0$, as $n \to \infty$.

Then, according to the famous *Optional Stopping Theorem*, we have

$$E(Y_N) = E(Y_1).$$

A proof of this theorem is given in many books; see, for example, Grimmett and Stirzaker (1992). Condition (c) may sometimes be difficult to check.

Example 2. The ruin problem

We shall return to the ruin problem already discussed in Section 1.5. Let A and B have initial capitals a and b and assume that they toss a fair coin and win or lose 1 unit until one of them is ruined. Player A's profit after n rounds can be written

$$Y_n = X_1 + \cdots + X_n,$$

where the X's assume the values 1 and -1 with the same probability $\frac{1}{2}$. By Example 1 the sequence $\{Y_n\}$ is a martingale. Let N be the number of rounds until one player is ruined; here N is the smallest integer n such that $Y_n = -a$ or $Y_n = b$ (in the first case A is ruined, and in the second case B is ruined). Let p be the probability that A is ruined. By the Optional Stopping Theorem (which is applicable here) we have

$$E(Y_N) = E(Y_1) = E(X_1) = 0.$$

But, on the other hand, it is clear that

$$E(Y_N) = -ap + b(1 - p).$$

Hence we obtain

$$p = \frac{b}{a + b},$$

in agreement with what we found in Section 1.5.

16.3 Wald's equation

Abraham Wald (1902–1950) is a great name in mathematical statistics. He is the father of sequential analysis and decision theory.

Wald's equation may be regarded as a by-product of sequential analysis. It looms in the background in Wald's classical book on sequential analysis from 1947, but was known already in 1944.

Let $X_1, X_2, \ldots$ be iid rv's with $E(|X_i|) < \infty$. Collect X's according to some stopping-rule and let N be a stopping time with $E(N) < \infty$. Set

$$Y_N = X_1 + \cdots + X_N. \qquad (1)$$

Hence Y_N is the sum of the X's when the sampling stops. *Wald's equation* states that

$$E(Y_N) = E(N)E(X).$$

In the special case when N is constant, the equation assumes the well-known form $E(Y_N) = N \cdot E(X)$.

Wald's equation has many applications. We shall give an application to gambling. Players A and B participate in a game consisting of several rounds. At each round, A wins 1 unit from B with probability $\frac{1}{2}$ and loses 1 unit to B with probability $\frac{1}{2}$. Before A begins the game, he chooses a *gambling strategy*. This implies that he selects a stopping-rule and hence also a stopping time N for the game.

During the game A's fortune will fluctuate. When the game ends, the total change of his fortune will be

$$Y_N = X_1 + \cdots + X_N,$$

where each X is 1 or -1 with the same probability $\frac{1}{2}$. We have, of course, $E(X) = 0$. All sensible gambling strategies are such that $E(N) < \infty$. Hence we may apply Wald's equation and obtain

$$E(Y_N) = E(N)E(X) = 0.$$

This is a marvellous result because of its generality. The gambler will not, on the average, increase or decrease his fortune, and a gambling system that helps the player to win in the long run does not exist. (If the expected gain in each round is positive, the conclusion is different.)

Last we shall prove Wald's equation using the theory of martingales. (The proof is only sketched.)

Let μ be the common mean of the X's defined at the beginning of the section. We find

$$E(Y_{n+1}|X_1,\ldots,X_n) = Y_n + E(X_{n+1}) = Y_n + \mu.$$

Set

$$Y_n' = \sum_{i=1}^{n}(X_i - \mu).$$

It is not very difficult to see that the sequence $\{Y_n'\}$ is a martingale. It is more difficult to show that the Optional Stopping Theorem is applicable; see the preceding section. Using this theorem we obtain the simple result

$$E(Y_N') = E(X_1 - \mu).$$

But this is equal to zero and so (noting that $Y_n' = Y_n - n\mu$) we have proved

$$E\left(\sum_{i=1}^{N} X_i\right) = E(N)\mu.$$

This is equivalent to Wald's equation.

16.4 Birth control

Suppose that in a family $P(\text{boy}) = P(\text{girl}) = \frac{1}{2}$ and that the sexes of the children are independent events. None of these assumptions are exactly true in real life. Two couples practice the following (admittedly rather queer) birth control: Couple 1 stops when a boy is immediately followed by another boy, couple 2 when a girl is immediately followed by a boy. This birth control is formally equivalent to a fair game with a stopping-rule. We now apply earlier results concerning patterns in random sequences; see Sections 1.4 and 14.2.

Denote by $E(N)$ the expected number of children in a family applying a given system of birth control. This quantity depends on the stopping-rule. Couple 1 will have $2 + 2^2 = 6$ children, and Couple 2 will have $2^2 = 4$ children, on the average.

How about the sex ratios in the families? Introduce zero–one rv's $X_1, X_2, \ldots$ such that X_i is 1 if the ith birth gives rise to a boy and 0 if it gives rise to a girl. Then

$$Y_n = X_1 + X_2 + \cdots + X_n,$$

is the number of boys in a family with n children. The X's are independent and have mean $\frac{1}{2}$. Now Y_N is the number of boys in a family applying a certain system of birth control. By Wald's equation in the preceding section we find

$$E(Y_N) = E(N)E(X) = \tfrac{1}{2}E(N).$$

Hence half of the children will be boys, on the average. As a consequence, the sex ratio is not affected by the birth control. This is an interesting and general result. Life would be different if it were not true!

16.5 The r-heads-in-advance game

In this section we give an application of the Optional Stopping Theorem and Wald's equation; see Sections 16.2 and 16.3.

Player A has a coin which shows heads with probability p_a and player B has a coin which shows heads with probability p_b. Each player tosses his coin until one of them has obtained r heads more than the the the other; he is the winner. We call this *the r-heads-in-advance game*. Find the probability P_A that A wins the game. Also find the expected number $E(N)$ of rounds until a winner is selected.

First we introduce some notation to be used in the sequel. Let the two sequences $X_1, X_2, \ldots$ and $Y_1, Y_2, \ldots$ be zero–one rv's which describe the outcome of the tosses. Set $X_i = 1$ if A's coin shows heads in round i, and $X_i = 0$ otherwise. Define Y_i analogously. Clearly, all the X's and Y's are iid with $P(X_i = 1) = 1 - P(X_i = 0) = p_a$ and $P(Y_i = 1) = 1 - P(Y_i = 0) = p_b$. Furthermore, let $Z_i = X_i - Y_i$ and

$$S_n = \sum_{i=1}^{n} Z_i.$$

The number N of rounds is the smallest integer n such that $|S_n| = r$. Moreover, $P_A = P(S_N = r)$ is the probability that A wins the game. Finally, define the constant λ by

$$\lambda = \frac{p_a(1 - p_b)}{p_b(1 - p_a)}. \tag{1}$$

Now consider the sequence $Z_1, Z_2, \ldots$ of iid rv's. The probability function of the Z's is given by

$$P(Z = -1) = P(X = 0, Y = 1) = (1 - p_a)p_b,$$
$$P(Z = 0) = P(X = Y) = p_a p_b + (1 - p_a)(1 - p_b),$$
$$P(Z = 1) = P(X = 1, Y = 0) = p_a(1 - p_b).$$

Hence we obtain

$$E(\lambda^{-Z}) = \lambda(1 - p_a)p_b + p_a p_b + (1 - p_a)(1 - p_b) + \frac{1}{\lambda}p_a(1 - p_b)$$
$$= p_a(1 - p_b) + p_a p_b + (1 - p_a)(1 - p_b) + (1 - p_a)p_b = 1.$$

Now consider the rv's

$$R_n = \lambda^{-S_n},$$

where $n = 1, 2, \ldots$. Note that $R_{n+1} = R_n \cdot \lambda^{-Z_{n+1}}$. We introduce the R's because they constitute a martingale with respect to the Z's. This is proved as follows: First, we have

$$E(R_{n+1}|Z_1, \ldots, Z_n) = E(R_n|Z_1, \ldots, Z_n) \cdot E(\lambda^{-Z_{n+1}}|Z_1, \ldots, Z_n).$$

Second, since R_n is determined by the rv's $Z_1, \ldots, Z_n$, and Z_{n+1} is independent of these rv's, we find

$$E(R_{n+1}|Z_1, \ldots, Z_n) = R_n \cdot E(\lambda^{-Z_{n+1}}) = R_n,$$

and so the martingale definition applies; see Section 16.2.

We now apply the Optional Stopping Theorem (one can prove that it is applicable here). The theorem shows that

$$E(R_N) = E(R_1).$$

Let us determine these two expectations. First, we have

$$E(R_1) = E(\lambda^{-Z_1}) = 1.$$

Second, we find

$$E(R_N) = E(\lambda^{-S_N})$$
$$= E(\lambda^{-S_N}|S_N = r) \cdot P_A + E(\lambda^{-S_N}|S_N = -r) \cdot (1 - P_A)$$
$$= \lambda^{-r} \cdot P_A + \lambda^r \cdot (1 - P_A) = \lambda^r + P_A \cdot (\lambda^{-r} - \lambda^r).$$

Hence the probability that A wins the game is given by

$$P_A = \frac{1 - \lambda^r}{\lambda^{-r} - \lambda^r} = \frac{\lambda^r}{1 + \lambda^r}. \tag{2}$$

Finally, noting that $E(Z) = p_a - p_b$ and using Wald's equation

$$E(S_N) = E(N)E(Z),$$

we obtain

$$E(N) = \frac{E(S_N)}{E(Z)} = \frac{rP_A - r(1 - P_A)}{p_a - p_b} = \frac{r(2P_A - 1)}{p_a - p_b}.$$

Inserting the value for P_A, we get for the expected number of rounds of the game

$$E(N) = \frac{r(\lambda^r - 1)}{(p_a - p_b)(\lambda^r + 1)}, \tag{3}$$

where λ is given in (1). In particular, if $p_a = p_b = p$, we find

$$E(N) = \frac{r^2}{2p(1 - p)}.$$

Can this problem, or parts of it, be solved for a similar game involving three players?

16.6   Patterns IV

In this section we shall generalize the results about patterns obtained in Section 14.3, using the elegant theory of martingales given in Section 16.2.

As before, we take digits from the set $\{1, 2, \ldots, m\}$ until one of n given patterns $S_1, S_2, \ldots, S_n$ appears. As usual, we are interested in the mean waiting time $E(N)$ and in the probabilities $\pi_1, \pi_2, \ldots, \pi_n$ that the procedure ends in $S_1, S_2, \ldots, S_n$, respectively. We now allow the digits to occur with probabilities $p_1, p_2, \ldots, p_m, \sum p_i = 1$, which may be different. Also, the patterns are allowed to have unequal lengths $k_1, k_2, \ldots, k_n$.

The overlaps between two patterns S_i and S_j are defined similarly as in the second paragraph of Section 14.3, with the modification that r now assumes the values $1, 2, \ldots, k$, where $k = \min(k_i, k_j)$. As a measure of the amount of overlap between S_i and S_j, in this order, we take

$$e_{ij} = \sum_{r=1}^{k} \frac{\epsilon_r(i, j)}{p_{c_1} \cdots p_{c_r}}, \tag{1}$$

where $\epsilon_r(i, j) = 1$ if there is a common part $c_1, \ldots, c_r$ of S_i and S_j, that is, if S_i ends with and S_j begins with $c_1, \ldots, c_r$, and $\epsilon_r(i, j) = 0$ otherwise.

Example 1

Digits are taken from the set $\{1, 2, 3\}$ with probabilities $1/2$, $1/3$, $1/6$, respectively, and the patterns are $S_1 = (1\ 2\ 2)$ and $S_2 = (2\ 3\ 1\ 2)$. Let us first look at the overlaps of S_2 with itself. We find

$$\epsilon_1(2, 2) = 1; \quad \epsilon_2(2, 2) = \epsilon_3(2, 2) = 0; \quad \epsilon_4(2, 2) = 1,$$

which gives the amount of overlap

$$e_{22} = \frac{1}{p_2} + \frac{1}{p_2 p_3 p_1 p_2} = 111.$$

Similarly we find

$$e_{11} = \frac{1}{p_1 p_2 p_2} = 18; \quad e_{12} = \frac{1}{p_2} = 3; \quad e_{21} = \frac{1}{p_1 p_2} = 6.$$

We now state the main result of the section:

The stopping probabilities $\pi_1, \pi_2, \ldots, \pi_n$ and the expected waiting time $E(N)$ are obtained by solving the system of equations

$$\sum_{j=1}^{n} e_{ji} \pi_j = E(N), \quad i = 1, \ldots, n,$$

$$\sum_{j=1}^{n} \pi_j = 1. \tag{2}$$

Example 1 (continued)

The system of equations (2) becomes in this case

$$18\pi_1 + 6\pi_2 = E(N),$$
$$3\pi_1 + 111\pi_2 = E(N),$$
$$\pi_1 + \pi_2 = 1.$$

The solution is $\pi_1 = 7/8$, $\pi_2 = 1/8$ and $E(N) = 33/2$.

A reader mainly interested in applications of the basic result (2) can stop reading here. In the rest of the section we sketch an ingenious proof due to Li (1980), which shows what may happen in the rarefied atmosphere of advanced probability.

The proof of (2) can be divided into two parts.

(a) *First part*

Let N_j be the waiting time until the single pattern S_j appears. Moreover, let N_{ij} be the waiting time until S_j appears in a sequence beginning with S_i. For instance, if $S_1 = (1 \quad 2 \quad 2)$, $S_2 = (1 \quad 1)$ and the sequence is $1\,2\,2\,1\,2\,1\,1$, we have $N_{12} = 4$. The expectations of the rv's N_j and N_{ij} are simple to write down:

$$E(N_j) = e_{jj}; \qquad E(N_{ij}) = e_{jj} - e_{ij}. \tag{3}$$

However, these relations are not simple to derive; for this purpose we will use the theory of martingales developed in Section 16.2.

Let l be a non-negative integer and consider the amount of overlap [defined by (1)] between the following sequences 1 and 2:

1. The pattern S_i followed by a sequence $x_1, x_2, \ldots, x_l$ selected from $\{1, 2, \ldots, m\}$ with probabilities $p_1, p_2, \ldots, p_m$.
2. The pattern S_j.

Let us denote this amount by $e(l)$; since it varies with the x's, it is an rv.
Define

$$Y(l) = e(l) - l. \tag{4}$$

It can be shown that the sequence of rv's

$$Y[\min(l, N_{ij})],$$

where $l = 0, 1, \ldots$, is a martingale and that the Optional Stopping Theorem holds for Y and the stopping time N_{ij}. Applying this theorem we obtain the relation

$$E[Y(N_{ij})] = E[Y(0)]. \tag{5}$$

We shall determine the expectations in (5).

First, look at the second expectation. It follows from (4) and the definition of $e(l)$ that $E[Y(0)]$ is the amount of overlap of S_i with S_j (in this order). That is, we have

$$E[Y(0)] = e_{ij}.$$

Second, look at the first expectation. Taking $l = N_{ij}$ in (4) we obtain

$$E[Y(N_{ij})] = E[e(N_{ij})] - E(N_{ij}). \tag{6}$$

Now $e(N_{ij})$ is the amount of overlap between a sequence, which begins with S_i and ends with S_j, and S_j. Hence the expectation of $e(N_{ij})$ is the amount of overlap of S_j with itself, and so we have proved

$$E[Y(N_{ij})] = e_{jj} - E(N_{ij}). \tag{7}$$

Inserting (6) and (7) into (5), we get the second expression (3). The first expression in (3) is a special case of the second; consequently, the proof of (3) is finished.

However, the proof is very incomplete: we have not verified the martingale property upon which the proof hinges, nor have we shown that green light can be given for applying the Optional Stopping Theorem.

(b) *Second part*

Let us rewrite the waiting time N_i for S_i. Remembering that N is the waiting time for the first of $S_1, \ldots, S_n$ and that $\pi_j = P(N = N_j)$, we obtain

$$E(N_i) = E(N) + E(N_i - N)$$

$$= E(N) + \sum_{j=1}^{n} E(N_i - N | N = N_j)\pi_j. \tag{8}$$

We know from the first expression in (3) that $E(N_i) = e_{ii}$. Furthermore, we obtain

$$E(N_i - N | N = N_j) = e_{ii} - e_{ji}, \tag{9}$$

for the expression on the left-hand side is the expected waiting time until S_i appears in a sequence starting with S_j, which by the second expression in (3) is equal to $e_{ii} - e_{ji}$. Inserting (9) into (8) and making a rearrangement we obtain (2), which is hereby proved.

16.7   Random permutation of 1's and (−1)'s

This section contains a beautiful, albeit not elementary, application of conditional probability. The proof we present is due to Dwass (1967).

(a) *The problem*

Consider a random permutation of n 1's and n (−1)'s. Denote this permutation by

$$U_1, U_2, \ldots, U_{2n}$$

and add $U_0 = 0$. Sum the U's successively:

$$W_0 = U_0,$$
$$W_1 = U_0 + U_1,$$

$$\vdots$$

$$W_{2n} = U_0 + U_1 + \cdots + U_{2n} \equiv 0.$$

Let N_n be the number of zeroes among the W's. Clearly, $N_n \geq 2$. We want to find the distribution of N_n. This distribution is of interest in nonparametric statistics, but here we regard the problem purely as one of discrete probability.

Example 1

Permute $1, 1, -1, -1$, which can be done in 6 ways, and insert 0 in front:

U_0	U_1	U_2	U_3	U_4	W_0	W_1	W_2	W_3	W_4
0	1	1	-1	-1	0	1	2	1	0
0	1	-1	1	-1	0	1	0	1	0
0	1	-1	-1	1	0	1	0	-1	0
0	-1	1	1	-1	0	-1	0	1	0
0	-1	1	-1	1	0	-1	0	-1	0
0	-1	-1	1	1	0	-1	-2	-1	0

Summing in the left-hand table from left to right, we obtain the right-hand table. It is found that N_2 is 2 with probability $2/6 = 1/3$ and 3 with probability $4/6 = 2/3$.

Returning to the general case, we shall prove the main result of the section:

$$P(N_n > k) = 2^k \binom{2n-k}{n-k} \bigg/ \binom{2n}{n} \qquad (1)$$

for $k = 1, 2, \ldots, n$. Using this expression it is easy to find the probability function of N_n. The proof of (1) consists of two parts, which we give in Subsections (b) and (c).

(b) A related random walk

Consider a random walk on the x-axis starting from the origin at $t = 0$ and moving at $t = 1, 2, \ldots$ one step to the right with probability $p < \frac{1}{2}$, or one step to the left with probability $q = 1 - p > \frac{1}{2}$.

Introduce independent rv's $X_1, X_2, \ldots$ such that the ith variable is 1 or -1 according to whether the ith step goes to the right or to the left. Set

$$S_1 = X_1,$$
$$S_2 = X_1 + X_2,$$
$$\vdots$$
$$S_{2n} = X_1 + X_2 + \cdots + X_{2n}.$$

We are interested in the event $S_{2n} = 0$. There are

$$\binom{2n}{n}$$

possible sequences $X_1, \ldots, X_{2n}$ leading to this event, each occurring with probability $p^n q^n$. The conditional probability of each such sequence, given that the event $S_{2n} = 0$ occurs, is equal to

$$\frac{p^n q^n}{\binom{2n}{n} p^n q^n} = \frac{1}{\binom{2n}{n}}.$$

Hence the conditional distribution of the X's, given $S_{2n} = 0$, assigns equal probabilities to the $\binom{2n}{n}$ possible sequences $X_1, \ldots, X_{2n}$, each consisting of n 1's and n (-1)'s. Some reflection discloses that this distribution is exactly the same as that of the U's in (a). This is attractive for lazy probabilists: instead of toiling with the dependent U's we may consider the independent X's and apply a conditioning procedure. We do this in the following way:

Let N, without subscript, denote the number of returns of the random walk to the origin when $n = 1, 2, \ldots$. This number is finite, with probability 1 because of the condition $p < \frac{1}{2}$; compare the exercise at the end of Section 10.7. Moreover, let T be the moment when the last return takes place. The conditioning procedure just described shows that

$$P(N_n > k) = P(N > k | T = 2n). \tag{2}$$

In words, the distribution of N_n is the same as the distribution of N given that $T = 2n$.

(c) *Main part of proof*

Now comes a 'technical' formula. We have from (2)

$$P(N > k) = \sum_{n=k}^{\infty} P(N > k | T = 2n) P(T = 2n)$$

$$= \sum_{n=k}^{\infty} P(N_n > k) P(T = 2n). \tag{3}$$

It is known from the end of Section 10.7 that the walk returns to the origin with probability $2p$. (Remember again that $p < \frac{1}{2}$.) Thereafter, a new return occurs independently with probability $2p$, and so on. As a consequence, we have

$$P(N > k) = (2p)^k \tag{4}$$

for $k = 0, 1, \ldots$. Also, the last return takes place at time $2n$ with probability

$$P(T = 2n) = \binom{2n}{n} p^n q^n (1 - 2p). \tag{5}$$

(The last factor is the probability that the walk never returns after time $2n$.) Combining (3), (4) and (5) we obtain an identity in p:

$$\frac{(2p)^k}{1 - 2p} = \sum_{n=k}^{\infty} \binom{2n}{n} (pq)^n P(N_n > k). \tag{6}$$

We now expand the left member in a power series of pq, and for that purpose use another of Dwass's ingenious devices: Consider once more the random walk on the x-axis and look at the first k steps. By a similar argument as that leading to (3) we find

$$p^k = P(X_1 = \cdots = X_k = 1)$$

$$= \sum_{n=k}^{\infty} P(X_1 = \cdots = X_k = 1 | T = 2n)P(T = 2n). \tag{7}$$

But

$$P(X_1 = \cdots = X_k = 1 | T = 2n) = \binom{k}{k}\binom{2n-k}{n-k} \Big/ \binom{2n}{n}, \tag{8}$$

since this is the probability that a sequence $X_1, \ldots, X_{2n}$ of n 1's and n (-1)'s starts with k 1's when all $\binom{2n}{n}$ sequences are equally probable.

Inserting (8) and expression (5) for $P(T = 2n)$ into (7), we obtain after a rearrangement the expansion

$$\frac{(2p)^k}{1-2p} = \sum_{n=k}^{\infty} 2^k \binom{2n-k}{n-k}(pq)^n.$$

Comparing this with (6) we obtain the expression for $P(N_n > k)$ given in (1), and the proof is finished. This is a rather long proof for such a simple formula as (1); however it is very instructive, and the same idea can be used in many other problems; see Dwass (1967).

17
Farewell problems

In this chapter we take farewell of our readers by presenting some problems
and snapshots connected with topics considered earlier in the book.

17.1 The first-to-r game II

(a) *Two players*

We change the first-to-r game considered in Section 15.8 to a situation
where A and B start with a white cards and b black cards, respectively.
One of the cards is removed at a time until one player has no cards left;
he is the winner. (The title is now false, but we persist in retaining that of
Section 15.8.)

First, let us find the expected number $E(N)$ of rounds. Using the
embedding technique described in Section 15.8, Subsection (c), we realize
that

$$E(N) = (a + b + 1) \int_0^1 (1 - x^a)(1 - x^b)\, dx = \frac{ab(a + b + 2)}{(a + 1)(b + 1)}.$$

Second, we have a new problem: What are the winning probabilities of the
players? Let P_A be A's winning probability. The answer is simple:

$$P_A = \frac{b}{a + b}. \tag{1}$$

We have found a short proof of this result which we are a little proud of.
Consider a random permutation of a ones and b zeroes. For example, if
$a = 2$ and $b = 4$, we may obtain 0 1 0 1 0 0. (Equivalently, we may draw
balls without replacement from an urn.)

The game between A and B can be performed by reading the permutation one step at a time from left to right until either all of the ones or all
of the zeros have been passed. In the example this happens after 4 steps;
the ones have then been passed, and A is the winner.

A simple observation provides a clue to the solution: player A wins
if and only if the permutation ends with a zero. Since this happens with
probability $b/(a + b)$, we obtain (1).

The argument just used is more or less obvious but should, strictly
speaking, be seen as a consequence of the exchangeability of a random
permutation; compare Section 16.1, especially Example 1.

(b) *Three players*

Players A, B and C start with a white, b black and c red cards, respectively. They play according to the usual rule until one of them has no cards left. What is A's winning probability P_A? The answer is

$$P_A = \frac{bc}{a+b+c}\left(\frac{1}{a+b} + \frac{1}{a+c}\right). \tag{2}$$

For the proof, we use exchangeability. Consider a random permutation of a 1's, b 2's and c 3's.

For example, if $a = 2$, $b = c = 3$, we may obtain the permutation 3 3 1 2 1 2 2 3. The game is equivalent to reading the permutation digit by digit from left to right until either all 1's, all 2's or all 3's have been passed. In this example, the 1's are passed after 5 steps, and A is the winner.

First we look at the remaining part of the permutation. It is perfectly general to say that A wins if and only if, after the last 1, there is at least one 2 and one 3.

Now reverse the permutation: 3 2 2 1 2 1 3 3. Reading this from left to right we understand that A wins the game if and only if either

1. the permutation begins with 2 and, thereafter, 3 appears before 1, or
2. the permutation begins with 3 and, thereafter, 2 appears before 1.

The former event occurs with probability

$$\frac{b}{a+b+c} \cdot \frac{c}{c+a}$$

and the latter with probability

$$\frac{c}{a+b+c} \cdot \frac{b}{b+a}.$$

Adding these two probabilities we obtain (2), which is hereby proved.

In the proof, we have used a property of exchangeable rv's with stopping times described in Section 16.1. For example, suppose that the reversed permutation begins with 2 so that there are a 1's, $b-1$ 2's and c 3's left. Now use the following stopping-rule: Go on reading until 3 or 1 appears and then stop. In view of what was said in Section 16.1, the stopping-rule does not affect what happens afterwards, and so the next digit is 3 or 1 with the unconditional probabilities $c/(c+a)$ and $a/(c+a)$, respectively.

Let us append a problem for the reader. Suppose that, after one player has won the game, the game continues until one of the two remaining players has no cards left; he comes in second. Show that A comes in second with probability

$$\frac{a}{a+b+c}\left(\frac{b}{a+c}+\frac{c}{a+b}\right)$$

and third with probability

$$\frac{a}{a+b+c}.$$

17.2   Random walk on a chessboard

A king is placed somewhere on an empty chessboard and performs a random walk. At each step, one of the possible moves is chosen with the same probability; see Figure 1. For each of the 64 squares, find the long run probability that the king visits the square. We will solve this problem using the theory of Markov chains; see Chapter 13.

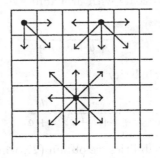

Fig. 1. Possible moves for king.

The chain has 64 states $E_1, \ldots, E_{64}$, one for each square. The number of possible moves from the state E_i is denoted by d_i. In the top left quadrant these numbers are

$$
\begin{array}{cccc}
3 & 5 & 5 & 5 \\
5 & 8 & 8 & 8 \\
5 & 8 & 8 & 8 \\
5 & 8 & 8 & 8 \\
\end{array}
$$

and the numbers for the other quadrants are obtained by symmetry considerations.

In the ith row of the 64×64 transition matris $P = (p_{ij})$, there are d_i elements equal to $1/d_i$, the other elements being zero.

Clearly we have

$$P(E_{i_1} \to E_{i_2} \to E_{i_3} \to E_{i_1}) = P(E_{i_1} \to E_{i_3} \to E_{i_2} \to E_{i_1})$$

and similarly for all other possible loops. Hence by Kolmogorov's criterion given in Section 13.8 the chain is reversible. By (1) of that section we have, for all elements of the asymptotic distribution, that $\pi_j p_{jk} = \pi_k p_{kj}$. Since for adjacent states we have $p_{jk} = 1/d_j$, this leads to the relation

$$\frac{\pi_j}{\pi_k} = \frac{d_j}{d_k}.$$

Thus the elements of the asymptotic distribution are given by

$$\pi_j = d_j \Big/ \sum_{k=1}^{64} d_k.$$

For example, the squares in the corners of the chessboard have long run probabilities $3/420$ and those in the central part $8/420$; compare Figure 1.

17.3   Game with disaster

The following problem appeared in 1982 in the Dutch journal *Statistica Neerlandica*: On a symmetric roulette with the numbers $0, 1, \ldots, m-1$ one plays until all the numbers $1, 2, \ldots, k$ have appeared at least once, with the restriction that no 0 has occurred. If a 0 occurs before $1, 2, \ldots, k$ all have appeared, the procedure starts again from the beginning. Find the expected duration of the game, that is, the number of plays performed when the game stops.

This problem describes a special case of the following general situation: Let $X_1, X_2, \ldots$ be a sequence of integer-valued iid rv's. Furthermore, let $Y_1, Y_2, \ldots$ be a sequence of iid rv's with common geometric probability function pq^{k-1}, where $k = 1, 2, \ldots$ and $q = 1 - p$. The X's and the Y's are assumed to be independent. Set $Z_i = \min(X_i, Y_i)$. Collect pairs $(X_1, Y_1), (X_2, Y_2), \ldots$ until for the first time the X-variable is the smallest in the pair; then stop. Let N be the number of pairs collected. Set

$$S = Z_1 + Z_2 + \cdots + Z_N.$$

We want to determine the expected value of S.

The special case in which we are interested is obtained as follows:

Extend the ith round until all the numbers $1, 2, \ldots, m-1$ have appeared. Define

$Y_i =$ number of plays finished when the first 0 appeared,

$X_i =$ number of plays finished when all the numbers $1, 2, \ldots, k$ have appeared, but excluding the plays resulting in 0.

Note that the number of plays actually performed in round no. i is $Z_i = \min(X_i, Y_i)$, and that X_i and Y_i are independent. Note also that S represents the number of plays in the entire game. For example, if $m = 3, k = 2$, the first round results in 1 1 0 and the first extended round in 1 1 0 1 0 2, we have $Y_1 = 3$, $X_1 = 4$. (Note also that we here have $Y_2 = 2$, $X_2 = 3$.)

We now solve the general problem. By Wald's equation in Section 16.3 we have

$$E(S) = E(N)E(Z), \tag{1}$$

where $Z = \min(X, Y)$. Hence we need the expectations of N and Z.

First, it is seen that N has a geometric distribution with probability function $\pi(1 - \pi)^{k-1}$ for $k = 1, 2, \ldots$, where $\pi = P(X < Y)$. Hence

$$E(N) = \frac{1}{\pi}. \tag{2}$$

Second, because of independence and the definition of Z we have

$$P(Z \geq k) = P(X \geq k)P(Y \geq k).$$

Using this relation we obtain successively

$$E(Z) = 1 \cdot P(Z = 1) + 2 \cdot P(Z = 2) + \cdots = \sum_{k=1}^{\infty} P(Z \geq k)$$

$$= \sum_{k=1}^{\infty} P(X \geq k)P(Y \geq k) = \sum_{k=1}^{\infty} P(X \geq k)q^{k-1}$$

$$= \frac{1}{p}\sum_{k=1}^{\infty} P(X \geq k)pq^{k-1} = \frac{1}{p}\sum_{k=1}^{\infty} P(X \geq k)P(Y = k)$$

$$= \frac{1}{p}P(X \geq Y) = \frac{1}{p}[1 - P(X < Y)].$$

Hence

$$E(Z) = \frac{1-\pi}{p}. \tag{3}$$

Inserting (2) and (3) into (1) we obtain the final result

$$E(S) = \frac{1-\pi}{\pi p}. \tag{4}$$

In the roulette problem, we have $p = 1/m$. Furthermore, π is the probability that 0 is the last number to appear among $0, 1, \ldots, k$; by symmetry we have $\pi = 1/(k + 1)$. Thus (4) yields

$$E(S) = mk.$$

Finally, we shall explain the title of the section. Consider a game consisting of several rounds with normal lengths $X_1, X_2, \ldots$. A round may be stopped abruptly by a disaster after Y plays (a die breaks into two parts, the casino burns, etc.). The probability is $1 - \pi$ that a round ends in this way, and the probability that a single play ends in a disaster is p. We are interested in the number of plays required until the game stops with a complete round without disaster. In the roulette problem, we want to gamble until the numbers $1, 2, \ldots, k$ have all appeared, and the occurrence of 0 is a disaster.

Further information about games with disasters is given by Janson (1985).

17.4   A rendezvous problem

Two units (people, particles etc.) move in a system of communication channels (roads, streets, edges in a graph, etc.) until they meet. Channels at forks are generally chosen by chance. Which strategy should the units choose in order to meet as soon as possible? The time until they meet is generally an rv, the expected value of which we denote by μ. We want to minimize μ. The units start at the same time, each unit from its own starting point, and they move at the same pace.

We call problems of the type sketched above *rendezvous problems*. We will analyse one such problem.

In a town there are four blocks of houses separated by streets; see Figure 1. At the start, A stands at corner $(0, 0)$ and B at corner $(2, 2)$. It takes the time $1/4$ to walk along a block and hence at least the time 1 to go from $(0, 0)$ to $(2, 2)$ or vice versa.

It is convenient to introduce the following terminology: A person is said to walk towards $(2, 2)$ if he always walks in the direction indicated by the arrows in Figure 2. He is said to walk towards $(0, 0)$ if he walks in the opposite direction.

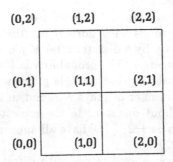

Fig. 1. Four blocks separated by streets.

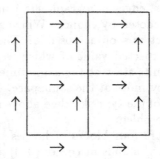

Fig. 2. Walking towards (2,2).

If a person walks towards $(2,2)$, say, and there are two directions to choose between on a street corner, he selects a road at random. A person walking towards $(0,0)$ does the same. Let us compare two strategies.

(a) Strategy 1

In the first period, A walks towards $(2,2)$ and B towards $(0,0)$. They may meet when they have passed two blocks, that is, after time $\frac{1}{2}$. If they do not meet after this time, A goes towards $(0,0)$ and B goes towards $(2,2)$ for an additional time $\frac{1}{2}$. A new period then begins, and so on.

The probability that the people meet after time $\frac{1}{2}$ at $(0,2)$, $(1,1)$ or $(2,0)$ is $1/16$, $1/4$ and $1/16$, respectively. If they do not meet, it takes the additional mean time $\frac{1}{2} + \mu$ until they meet. Hence we have the relation

$$\mu = \frac{3}{8} \cdot \frac{1}{2} + \frac{5}{8} \cdot (1 + \mu),$$

which has the solution $\mu = 13/6 \approx 2.1667$.

(b) *Strategy 2*

First A walks as before for the time $\frac{1}{2}$ towards $(2,2)$ and B walks towards $(0,0)$. If they do not meet, each person decides by tossing a coin whether to walk towards $(0,0)$ or $(2,2)$. During this part of the walk, they may meet after an additional time $\frac{1}{4}$ or $\frac{1}{2}$, or not at all. In the third case they have walked for the total time 1, and a new period begins, and so on.

It is now somewhat more complicated to find μ than before. After time $\frac{1}{2}$ there are three main cases and some subcases:

1. The people meet at $(0,2)$, $(1,1)$ or $(2,0)$. As we saw earlier, this happens with probability $3/8$.
2. One person arrives at $(0,2)$, the other at $(2,0)$. This happens with probability $1/8$. After an additional time $1/2$ they may either meet at one of the starting points or not meet there; the probabilities of these subcases are $1/2$. If they do not meet, a new period begins.
3. One person arrives at $(1,1)$ and the other at $(2,0)$ or $(0,2)$. This happens with probability $1/2$. For the remaining part of the period there are three subcases:
   a. The people meet after an additional time $1/4$. This happens with probability $1/4$. [For example, if after half the period one person is at $(1,1)$ and the other at $(0,2)$, they may meet at $(0,1)$ or $(1,2)$.]
   b. The people meet after an additional time $1/2$. This also happens with probability $1/4$. (For example, if after half the period A is at $(1,1)$ and B is at $(0,2)$, A may walk to $(2,2)$ via $(2,1)$ and B may also walk to $(2,2)$, or A may walk to $(0,0)$ via $(1,0)$ and B may also walk to $(0,0)$).
   c. The people do not meet. This happens with probability $1/2$. A new period then begins.

Combining these cases and subcases we obtain the relation

$$\mu = \frac{3}{8} \cdot \frac{1}{2} + \frac{1}{8} \cdot \left[\frac{1}{2} \cdot 1 + \frac{1}{2} \cdot (1 + \mu) \right]$$
$$+ \frac{1}{2} \cdot \left[\frac{1}{4} \cdot \frac{3}{4} + \frac{1}{4} \cdot 1 + \frac{1}{2} \cdot (1 + \mu) \right].$$

Solving this equation in μ, we find $\mu = 25/22 \approx 1.1364$. Thus Strategy 2 is much better than Strategy 1.

New kinds of problems arise if we want to determine the strategy that gives the smallest μ among all possible strategies.

Here is a problem for the reader: Consider a cube with corners at $(0,0,0), (0,0,1), \ldots, (1,1,1)$ in a three-dimensional coordinate system. It takes the time $1/3$ for a particle to move along one side of the cube. At the start, particle 1 is at $(0,0,0)$ and particle 2 at $(1,1,1)$.

In the first period, particle 1 moves toward $(1,1,1)$ and particle 2 moves toward $(0,0,0)$ for time $1/2$. (They then move along a whole edge for time $1/3$ and along half an edge for time $1/6$.) The particles may then meet. If they do not meet after this time, particle 1 goes toward $(0,0,0)$ and particle 2 toward $(1,1,1)$ for time $1/2$; then a new period begins. Show that $\mu = 11/2$.

A rendezvous problem of another type than the one considered in this section has been discussed by Anderson and Weber (1990).

17.5   Modified coin-tossing

A coin shows heads ($=1$) with probability p and tails ($=0$) with probability $q = 1 - p$. The coin is tossed repeatedly and the result is written down. However, a 1 followed by a 1 is forbidden, so when a 1 is obtained, 0 is written in the protocol as the result of the next toss. Hence, a sequence like 1011011 is forbidden, but 1010010 is allowed.

An application of such modified coin-tossing is the following: Players A and B toss a coin several times. A wins if 1 appears and B wins if 0 appears. Player A says to B: 'I want to be generous. When I have won a round, we say that you have won the next one.'

In the sequel, an allowable sequence of length n is called a *10–sequence* (1 is always followed by 0). We are interested in the number of 10–sequences and in the number of 1's in such a sequence.

(a) *Number of 10–sequences*

We need the famous *Fibonacci numbers* $F_0, F_1, \ldots$ where $F_0 = 0$, $F_1 = 1$, $F_2 = 1$, $F_3 = 2$, $F_4 = 3$, $F_5 = 5$ and, generally, $F_{k+2} = F_{k+1} + F_k$; each number being the sum of the two preceding ones. Many properties of these numbers are described by Graham, Knuth and Patashnik (1989, p. 276). *Leonardo Fibonacci* (c. 1180–1250) was an Italian mathematician who introduced his numbers in 1202 in his famous book *Liber Abaci* (The Book of the Abacus).

The nth Fibonacci number can be represented as

$$F_n = \frac{1}{\sqrt{5}}\left[\left(\frac{1+\sqrt{5}}{2}\right)^n - \left(\frac{1-\sqrt{5}}{2}\right)^n\right]. \tag{1}$$

This may be shown in different ways, for example by induction.

Let now c_n be the number of 10–sequences. For $n = 1$ there are 2 sequences 0 and 1, for $n = 2$ there are 3 sequences 00, 01, 10 and for $n = 3$ there are 5 sequences 000, 001, 010, 100, 101. These numbers of sequences satisfy $5 = 2 + 3$, which is no coincidence. In fact, the c's constitute for $n = 1, 2, \ldots$ a Fibonacci sequence $2, 3, 5, 8, 13, \ldots$; that is, $c_n = F_{n+2}$. The proof is simple and is left to the reader.

What has been said here is not probability but is included in order to show the nice relationship to the Fibonacci numbers. Besides, the result can be used for solving the following problem in probability: Toss a fair coin n times. Show that the probability P_n that a 10–sequence is obtained is given by

$$P_n = \frac{c_n}{2^n} = \frac{1}{2^n} \cdot \frac{1}{\sqrt{5}} \cdot \left[\left(\frac{1+\sqrt{5}}{2}\right)^{n+2} - \left(\frac{1-\sqrt{5}}{2}\right)^{n+2}\right].$$

(b) *Expectation of the number of 1's*

Let X_n be the number of 1's in a 10–sequence, when throwing a biased coin. We want to determine the expectation $\mu_n = E(X_n)$.

Write X_n as a sum

$$X_n = U_1 + U_2 + \cdots + U_n,$$

where U_i is the digit written at the ith position in the protocol. The U's are dependent zero–one rv's.

Set $\alpha_i = E(U_i)$ for $i = 1, \ldots, n$. We successively find

$$\alpha_1 = P(U_1 = 1) = p,$$
$$\alpha_2 = P(U_2 = 1) = P(U_1 = 0)p = (1 - p)p = p - p^2,$$
$$\alpha_3 = P(U_3 = 1) = P(U_2 = 0)p = (1 - p + p^2)p = p - p^2 + p^3,$$

and generally

$$\alpha_n = (1 - \alpha_{n-1})p.$$

Adding the α's, we find that the first three μ's become p, $2p - p^2$ and $3p - 2p^2 + p^3$. Generally, we have

$$\mu_n = np - (n - 1)p^2 + \cdots + (-1)^{n-1}p^n.$$

It may be shown, for example by induction, that

$$\mu_n = \frac{np}{1+p} + \left(\frac{p}{1+p}\right)^2 [1 - (-p)^n].$$

(c) Probability function of the number of 1's

We shall derive the probability function $P(X_n = k)$, where the range of variation will be given below. We distinguish between two cases:

1. The n-sequence ends with 0. The probability that we get k 1's, none of which are situated at the end, is the 'binomial probability'

$$A_{nk} = \binom{n-k}{k} p^k q^{n-2k}.$$

2. The n-sequence ends with 1. Again we have a binomial probability, but this time multiplied by p for the last 1:

$$B_{nk} = \binom{n-k}{k-1} p^{k-1} q^{n-2k+1} \cdot p.$$

Adding A_{nk} and B_{nk} we obtain $P(X_n = k)$ (using the convention that $A_{nk} = 0$ if $n < 2k$ and $B_{n0} = 0$). The range of variation is $k = 0, 1, \ldots, n/2$ if n is even, and $k = 0, 1, \ldots, (n+1)/2$ if n is odd.

We suggest several problems for the interested reader:

First, analyse the two-state Markov chain generated by the modified coin tosses discussed in this section.

Second, change the rules of the game given as an illustration at the beginning of the section in the following way: only when A has won two rounds in succession does he 'give' the next round to B.

Third, A and B may be equally generous: When one of them has won two rounds in succession, the other player 'gets' the next round.

Fourth, three players A, B and C participate and win a round with the same probability 1/3. When A or B win a round, they give the next round to C. The numbers of allowable sequences of size $1, 2, \ldots$ can even in this case be found recursively. Find the recursive relation.

Reference: Horibe (1990).

17.6   Palindromes

This section is devoted to a problem, which, we are almost sure, has not been discussed before in the literature. We consider a fictitious reader of our book who one day says to himself:

'I have read a lot about tosses of coins and throws of dice in this book. Since I have always been interested in palindromic words such as "mum" or "madam", I sometimes look for palindromic sequences such as 12221 when tossing a coin or 165424561 when throwing a die. It seems fun to imagine the following situation: Toss the coin, or throw the die, until the whole sequence of results, counted from the beginning, becomes palindromic. What is the probaility that this ever happens? This problem may seem a bit crazy, but it appeals to me, and I would be happy if it were solved.

'It is easy to prove that a coin eventually produces a palindromic sequence, but a die seems reluctant to do so. For example, if I toss a coin repeatedly and call heads 1 and tails 2, the result becomes perhaps 1, 12, 122, 1222, 12221, the last sequence being a palindrome. But if I throw a die, I may obtain 2, 21, 216, 2164, 21644, 216443, and there is little hope that a palindrome will ever turn up. However, the probability is not zero since, for example, the six following throws may result in 344612.

'I now state the following general problem: The random digits x_1, $x_1 x_2$, $x_1 x_2 x_3$, ... are collected from the set $\{1, 2, ..., m\}$. Look at these sequences of increasing lengths. Find the probability P that a palindrome turns up sooner or later.'

This is a harder problem than might be thought at first sight.

We use the term 'digit' for the number taken from the set, although it is not entirely adequate if $m \geq 10$. The results also hold if the set consists of other objects than digits. For example, we may select letters at random from the set $\{A, B, ..., Z\}$.

(a) Palindromes in sequences of given length

We begin with a simpler problem. Let us determine the probability p_n that a given number $n \geq 2$ of random digits $x_1 x_2 ... x_n$ taken from the set $\{1, 2, ..., m\}$ form a palindrome. When n is even, so that $n = 2k$ say, the second half of the palindrome is determined by the first half. When n is odd, so that $n = 2k + 1$, the digit in the middle can be any of the m possible digits; if it is excluded, the situation is the same as in the previous case. Hence we conclude that when $n = 2k$ or $n = 2k + 1$, we have

$$p_{2k} = p_{2k+1} = p^k, \tag{1}$$

where $p = 1/m$ and $k = 1, 2, ...$.

(b) An upper bound for P

Let r_k be the probability that a palindrome is obtained for the first time when k digits have been collected. The probability P can be written

$$P = r_2 + r_3 + \cdots . \tag{2}$$

Before we discuss the probabilities r_j, we shall derive an upper limit for P. For any $j \geq 2$ we have, clearly, the inequality

$$r_j \leq p_j, \tag{3}$$

where p_j is given by (1).

As an illustration, consider the binary case $m = 2$. If j is given, say equal to 3, there are four possible palindromes 111, 121, 212, 222; hence we have $p_3 = 4/2^3 = 1/2$; this also follows from (1). On the other hand, if we collect digits one at a time, the possible palindromes of length 3 are reduced to 121, 212, hence from 4 to 2, and so $r_3 = 2/2^3 = 1/4$.

From (2) and (3) we obtain the inequality

$$P \leq p_2 + p_3 + \cdots. \tag{4}$$

Using (1) we obtain after a summation

$$P \leq \frac{2p}{1-p},$$

where $p = 1/m$. This inequality is rather coarse; it shows that the probability of a palindrome is less than 1 when $m \leq 4$. For throws of a die we obtain $P \leq 2/5 = 0.4$.

(c) *Calculation of the r's*

The probabilities $r_2, r_3, \ldots, r_n$ in (2) can be determined, using the recursive relations

$$r_{2k} = \frac{1}{m^k}(1 - r_2 - r_3 - \cdots - r_k),$$

$$r_{2k+1} = \frac{1}{m^k}(1 - r_2 - r_3 - \cdots - r_{k+1}). \tag{5}$$

The recursion starts with $r_2 = 1/m$. A proof of (5) is given in the appendix of this section. For numerical calculations, the following equivalent relations are convenient:

$$r_{2k} = \frac{1}{m}r_{2k-1},$$

$$r_{2k+1} = r_{2k} - \frac{1}{m^k}r_{k+1}. \tag{6}$$

Example

Let us consider the special case $m = 3$. From (5) or (6) we obtain the following values successively:

k	2	3	4	5	6	7
r_k	$\frac{1}{3}$	$\frac{2}{9}$	$\frac{2}{27}$	$\frac{4}{81}$	$\frac{4}{243}$	$\frac{10}{729}$

Addition of the six first r's gives

$$P = \frac{517}{729} \approx 0.7092.$$

This is an approximate value of the probability that a palindrome is obtained when there are three possible values 1, 2 and 3 in the set. (Similarly, it can be shown that, if a die is thrown, the probability is 0.356 that a palindrome turns up sooner or later.)

Here is a problem for the reader: In a 4×4 matrix the 16 elements independently assume values 1 and 2 with the same probability. Let X be the number of rows and columns that are palindromic. Show that

a. the mean and variance of X are, respectively, 2 and 3/2,
b. the rv X attains its largest possible value 8 with probability $(\frac{1}{2})^{12}$,
c. the probability that X is exactly 1 is 1/4.

[Hint: use a computer. We have also solved the third of these problems with pen and paper alone.]

Appendix

This appendix contains a proof of the recursive relations (5) above. The proof is divided into two parts, the first containing a mathematical statement required for the probabilistic argument of the proof in the second part. It is recommended that the reader begin with the second part and consult the first when necessary.

(a) Partial palindromes

We need a new term which we invented: *partial* palindrome; it will be introduced by means of an example:

The palindrome 12121 starts with the palindrome 121, which is a partial palindrome of 12121. Only palindromes at the beginning are considered. The palindrome 1212121 has two partial palindromes, 121 and 12121.

A palindrome containing one or more partial palindromes is said to be a *Type 1* palindrome. A palindrome without partial palindromes is said to be a *Type 2* palindrome. Example: 1212121 is a Type 1 palindrome and 1222221 a Type 2 palindrome.

Consider a palindrome A of length n taken from the set $\{1, 2, \ldots, m\}$, where $n = 2k$ or $n = 2k+1$ and $k = 2, 3, \ldots$. Suppose that A has a partial palindrome B of length $n - e$, where

$$0 < e < k.$$

It then holds that, besides B, the palindrome A has one or more partial palindromes of lengths

$$n - 2e, n - 3e, \ldots, \Delta,$$

where $2 \le \Delta \le e + 1$. Here Δ is the length of the shortest partial palindrome.

The proof will not be given here. Let us illustrate the statement with an example. Consider the palindrome

$$A = 11211211211,$$

which has odd length, $n = 11$. It has a partial palindrome

$$B = 11211211$$

of length $11 - 3 = 8$. By the property just demonstrated, A has also partial palindromes of lengths $11 - 2 \cdot 3 = 5$ and $11 - 3 \cdot 3 = 2$; this is obviously true.

(b) *Proof of the recursive relations*

Let us recall the situation described in detail at the beginning of the section. We select digits x_1, x_1x_2, $x_1x_2x_3$, $\ldots$ from the set $\{1, 2, \ldots, m\}$ until a palindrome turns up (if it does!) and want to find the probability r_n that this happens for the first time when n digits have been collected. Such a palindrome is necessarily of Type 2, that is, it has no partial palindromes. (If it had, the procedure would have stopped earlier.)

We want to prove the relations

$$r_{2k} = \frac{1}{m^k}(1 - r_2 - r_3 - \cdots - r_k),$$

$$r_{2k+1} = \frac{1}{m^k}(1 - r_2 - r_3 - \cdots - r_{k+1}). \tag{A1}$$

Consider all palindromes of given even length $n = 2k$ or of given odd length $n = 2k + 1$. Earlier we introduced the notation p_{2k} and p_{2k+1} for the probabilities that a such a palindrome is obtained and know that

$$p_{2k} = p_{2k+1} = p^k, \tag{A2}$$

where $p = 1/m$ and $k = 1, 2 \ldots$.

Let q_{2k} and q_{2k+1} be the probabilities that a Type 1 palindrome is obtained. Since the q's and the r's add up to the p's in $(A2)$, we have

$$r_{2k} = p^k - q_{2k}; \qquad r_{2k+1} = p^k - q_{2k+1}. \tag{A3}$$

In order to find the r's we determine the q's. We consider only the case of even lengths, $n = 2k$, since the odd case is handled quite similarly.

Let A_{2k} be a Type 1 palindrome of length $2k$. It has a smallest partial palindrome D_j, say, of length j; this palindrome is, of course, of Type 2. The possible values of j are only $2, 3, \ldots, k$; this is a consequence of the statement in (a) and is proved as follows:

Suppose, for example, that $j = k+1$ were a possible value so that A_{2k} had a smallest partial palindrome B of length $k+1$, that is, of length $2k-e$ with $e = k-1$. By the statement in (a), A_{2k} would then also have a partial palindrome C of length $2k - 2e = 2$; this contradicts the assumption that B is smallest.

It follows from what has been said above that a Type 1 palindrome $x_1 \ldots x_k x_k \ldots x_1$ is obtained if one of the disjoint events $E_2, E_3, \ldots, E_k$ occurs, where E_j is a 'threefold' event: first, $x_1 \ldots x_j$ is a Type 2 palindrome; second, $x_{j+1} \ldots x_k$ are any digits; third, the last k digits are $x_k \ldots x_1$. The probability $P(E_j)$ of this compound event is readily seen to be r_j/m^k. Summing over j, we obtain the total probability q_{2k} of a Type 1 palindrome of length $2k$ in the form

$$q_{2k} = \frac{1}{m^k}(r_2 + r_3 + \cdots + r_k).$$

Finally, using the first part of $(A3)$, we obtain the first part of $(A1)$.

As we have already said, the second part of $(A1)$ is proved in a similar way.

References

ALDOUS, D. J. (1991). Threshold limits for cover times. *Journal of Theoretical Probability*, **4**, 197–211.

ALDOUS, D. J. AND DIACONIS, P. (1986). Shuffling cards and stopping times. *The American Mathematical Monthly*, **93**, 333–348.

ALELIUNAS, R., KARP, R. M., LIPTON, R. J., LOVÁSZ, L., AND RACK-OFF, C. (1979). Random walks, universal traversal sequences, and the complexity of maze problems. In *20th Annual Symposium on Foundations of Computer Science*, October 1979, 218–223.

ALON, N. AND SPENCER, J. (1992). *The Probabilistic Method*. Wiley, New York.

ANDERSON, E. J. AND WEBER, R. R. (1990). The rendezvous problem on discrete locations. *Journal of Applied Probability*, **28**, 839–851.

ARNOLD, B. C., BALAKRISHNAN, N., AND NAGARAJA, H. N. (1992). *A First Course in Order Statistics*. Wiley, New York.

BARBOUR, A. D. AND HALL, P. (1984). On the rate of Poisson convergence. *Mathematical Proceedings of the Cambridge Philosophical Society*, **95**, 473–480.

BARBOUR, A. D., HOLST, L., AND JANSON, S. (1992). *Poisson Approximation*. Clarendon Press, Oxford.

BARTON, D. E. AND MALLOWS, C. L. (1965). Some aspects of the random sequence. *Annals of Mathematical Statistics*, **36**, 236–260.

BAYER, D. AND DIACONIS, P. (1992). Trailing the dovetail shuffle to its lair. *Annals of Applied Probability*, **2**, 294–313.

BAYES, T. (1764). An essay towards solving a problem in the doctrine of chances. *Philosophical Transactions*, **53**, 370–418. Reprinted 1970 in *Studies in the History of Statistics and Probability*, Vol. 1, edited by E. S. Pearson and M. Kendall. Griffin, London.

BERNOULLI, J. (1713). *Ars Conjectandi*. Reprinted in *Die Werke von Jakob Bernoulli*, Vol. 3 (1975). Birkhäuser, Basel.

BLOM, G. (1985). A simple property of exchangeable random variables. *The American Mathematical Monthly*, **92**, 491–492.

BLOM, G. (1989a). Mean transition times for the Ehrenfest urn model. *Advances in Applied Probability*, **21**, 479–480.

BLOM, G. (1989b). *Probability and Statistics: Theory and Applications.* Springer–Verlag, New York.

BLOM, G. AND HOLST, L. (1986). Random walks of ordered elements with applications. *The American Statistician,* **40**, 271–274.

BLOM, G. AND HOLST, L. (1989). Some properties of similar pairs. *Advances in Applied Probability,* **21**, 941–944.

BLOM, G. AND HOLST, L. (1991). Embedding procedures for discrete problems in probability. *The Mathematical Scientist,* **16**, 29–40.

BLOM, G. AND SANDELL, D. (1992). Cover times for random walks on graphs. *The Mathematical Scientist,* **17**, 111–119.

BLOM, G. AND THORBURN, D. (1982). How many random digits are required until given sequences are obtained? *Journal of Applied Probability,* **19**, 518–531.

BLOM, G., THORBURN, D., AND VESSEY, T. A. (1990). The distribution of the record position and its applications. *The American Statistician,* **44**, 151–153.

BOGART, K. P. AND DOYLE, P. G. (1986). Non-sexist solution of the ménage problem. *The American Mathematical Monthly,* **93**, 514–518.

BORTKEWITSCH, L. VON (1898). *Das Gesetz der kleinen Zahlen.* Teubner Verlag, Leipzig.

CARDANO, G. (1663). *Liber de Ludo Aleae.* English translation in ORE (1953).

CHEN, R. AND ZAME, A. (1979). On fair coin-tossing games. *Journal of Multivariate Analysis,* **9**, 150–156.

DALE, A. I. (1991). *A History of Inverse Probability.* Springer-Verlag, New York.

DAVID, F. N. AND BARTON, D. E. (1962). *Combinatorial Chance.* Griffin, London.

DAVID, H. A. (1981). *Order Statistics* (2nd ed). Wiley, New York.

DAWKINS, B. (1991). Siobhan's problem: the coupon collector revisited. *The American Statistician,* **45**, 76–82.

DOOB, J. L. (1953). *Stochastic Processes.* Wiley, New York.

DURRETT, R. (1991). *Probability: Theory and Examples.* Wadsworth & Brookes/Cole, Belmont, CA.

DUTKA, J. (1988). On the St. Petersburg Paradox. *Archives for History of Exact Sciences,* **39**, 13–39.

DWASS, M. (1967). Simple random walk and rank order statistics. *Annals of Mathematical Statistics*, **38**, 1042–1053.

EGGENBERGER, F. AND PÓLYA, G. (1923). Über die Statistik Verketteter Vorgänge. *Zeitschrift für Angewandthe Mathematik und Mechanik*, **3**, 279–289.

EHRENFEST, P. AND EHRENFEST, T. (1907). Über zwei bekannte Einwände gegen das Boltzmannsche H–Theorem. *Physikalische Zeitschrift*, **8**, 311–314.

Encyclopedia of Statistical Sciences (1982–1988), edited by S. Kotz and N. L. Johnson. Wiley, New York.

ENGEL, E. AND VENETOULIAS, A. (1991). Monty Hall's probability puzzle. *Chance*, **4**, 6–9.

FELLER, W. (1968). *An Introduction to Probability Theory and Its Applications*, Vol. I (3rd ed). Wiley, New York.

FELLER, W. (1971). *An Introduction to Probability Theory and Its Applications*, Vol. II (2nd ed). Wiley, New York.

FIBONACCI, L. (1202). *Liber Abaci* (1st ed). First edition now lost; second edition (1228) reprinted in *Scritti di Leonardo Pisano*, **1** (1857).

GLICK, N. (1978). Breaking records and breaking boards. *The American Mathematical Monthly*, **85**, 2–26.

GOLDIE, C. M. AND ROGERS, L. C. G. (1984). The k-record processes are i.i.d.. *Zeitschrift für Wahrscheinlichkeitstheorie und verwandte Gebiete*, **67**, 197–211.

GRAHAM, R. L., KNUTH, D. E., AND PATASHNIK, O. (1989). *Concrete Mathematics*. Addison-Wesley, Reading, MA.

GRIMMETT, G. R. AND STIRZAKER, D. R. (1992). *Probability and Random Processes* (2nd ed). Clarendon Press, Oxford.

HAIGHT, F. A. (1967). *Handbook of the Poisson Distribution*. Wiley, New York.

HALD, A. (1990). *A History of Probability & Statistics and Their Applications before 1750*. Wiley, New York.

HEATH, D. AND SUDDERTH, W. (1976). De Finetti's theorem on exchangeable variables. *The American Statistician*, **30**, 188–189.

HOLST, L. (1991). On the 'problème des ménages' from a probabilistic viewpoint. *Statistics & Probability Letters*, **11**, 225–231.

HORIBE, Y. (1990). Some notes on Fibonacci binary sequences. In *Applications of Fibonacci Numbers*. Kluwer Academic Publishers, Dordrecht, 155–160.

HUYGENS, C. (1657). *De Ratiociniis in Ludo Aleae*. Reprinted in *Oeuvres*, **14** (1920).

ISAACSON, D. L. AND MADSEN, R. W. (1976). *Markov Chains: Theory and Applications*. Wiley, New York.

JANSON, S. (1985). On waiting times in games with disasters. In *Contributions to Probability and Statistics in Honour of Gunnar Blom*, edited by J. Lanke and G. Lindgren. Department of Mathematical Statistics, University of Lund.

JOHNSON, N. L. AND KOTZ, S. (1969). *Distributions in Statistics: Discrete Distributions*. Houghton Mifflin, Boston.

JOHNSON, N. L. AND KOTZ, S. (1977). *Urn Models and Their Application*. Wiley, New York.

KALLENBERG, O. (1982). Characterizations and embedding properties in exchangeability. *Zeitschrift für Wahrscheinlichkeitstheorie und verwandte Gebiete*, **60**, 249–281.

KALLENBERG, O. (1985). Some surprises in finite gambling and its continuous time analogue. In *Contributions to Probability and Statistics in Honour of Gunnar Blom*, edited by J. Lanke and G. Lindgren. Department of Mathematical Statistics, University of Lund.

KELLY, F. P. (1979). *Reversibility and Stochastic Networks*. Wiley, New York.

KEMENY, J. G. AND SNELL, J. L. (1960). *Finite Markov Chains*. Van Nostrand, Princeton.

KNUTH, D. (1992). Two notes on notations. *The American Mathematical Monthly*, **99**, 403–422.

LAPLACE, P. S. (1812). *Théorie analytiques des probabilités*. Paris. Reprinted in *Oeuvres*, **7** (1886).

LI, S. R. (1980). A martingale approach to the study of occurrence of sequence patterns in repeated experiments. *Annals of Probability*, **8**, 1171–1176.

MAISTROV, L. E. (1974). *Probability Theory. A Historical Sketch*. Academic Press, New York.

MARTIN–LÖF, A. (1985). A limit theorem which clarifies the 'Petersburg paradox'. *Journal of Applied Probability*, **22**, 634–643.

MOIVRE, A. DE (1712). De Mensura Sortis. *Philosophical Transactions*, **27**, 213–264. Translated into English by B. McClintock (1984) in *International Statistical Review*, **52**, 237–262.

MOIVRE, A. DE (1756). *The Doctrine of Chances. The third edition, fuller, clearer, and more correct than the former.* Millar, London. Reprinted 1967 by Chelsea, New York.

MONTMORT, P. R. DE (1713). *Essay d'analyse sur les jeux de hazard* (2nd ed). Jacques Quillau, Paris. Reprinted 1980 by Chelsea, New York.

MOOD, A. M. (1940). The distribution theory of runs. *Annals of Mathematical Statistics*, **11**, 367–392.

MOOD, A. M. (1943). On the dependence of sampling inspection plans upon population distributions. *Annals of Mathematical Statistics*, **14**, 415–425.

MORAN, P. A. P. (1968). *An Introduction to Probability Theory.* Clarendon Press, Oxford. Reprinted, paperback edition, 1984.

MORGAN, J. P., CHAGANTY, N. R., DAHIYA, R. C., AND DOVIAK, M. J. (1991). Let's make a deal: the player's dilemma. *The American Statistician*, **45**, 284–289.

MOSTELLER, F. (1965). *Fifty Challenging Problems in Probability.* Addison–Wesley, Reading, MA. Reprinted, Dover, New York, 1987.

NAGARAJA, H. N. (1988). Record values and related statistics – a review. *Communications in Statistics – Theory and Methods*, **17**, 2223–2238.

NEVZOROV, V. B. (1987). Records. *Theory of Probability and Its Applications*, **32**, 201–228.

ORE, O. (1953). *Cardano the Gambling Scholar.* Princeton University Press, Princeton, NJ.

OWEN, G. (1968). *Game Theory.* Saunders, Philadelphia.

PALACIOS, J. L. (1990). On a result of Aleliunas et al. concerning random walks on graphs. *Probability in the Engineering and Informational Sciences*, **4**, 489–492.

PENNEY, W. (1969). Problem: penney-ante. *Journal of Recreational Mathematics*, **2**, 241.

POINCARÉ, H. (1912). *Calcul des probabilités.* Gauthier-Villars, Paris.

POISSON, S. D. (1837). *Recherches sur la probabilité des jugements en matière criminelle et en matière civile, précédées des règles générales du calcul des probabilités.* Bachelier, Paris.

PÓLYA, G. (1921). Über eine Aufgabe der Wahrscheinlichkeitsrechnung betreffend die Irrfahrt im Strassennetz. *Mathematische Annalen*, **84**, 149–160.

Pólya Picture Album: Encounters of a Mathematician (1987), edited by G. L. Alexanderson. Birkhäuser: Boston and Basel.

RÉNYI, A. (1969). *Briefe über die Wahrscheinlichkeit*. Birkhäuser, Basel.

ROSS, S. M. (1983). *Stochastic Processes*. Wiley, New York.

SAVANT, M. VOS (1990a). Ask Marilyn. *Parade Magazine*, Sept. 9, p. 15.

SAVANT, M. VOS (1990b). Ask Marilyn. *Parade Magazine*, Dec. 2, p. 25.

SAVANT, M. VOS (1991). Ask Marilyn. *Parade Magazine*, Febr. 17, p. 12.

SPITZER, F. (1964). *Principles of Random Walk*. Van Nostrand, Princeton.

STIGLER, S. M. (1986). *The History of Statistics*. Harvard University Press, Cambridge, MA.

TAKÁCS, L. (1962). A generalization of the ballot problem and its application in the theory of queues. *Journal of the American Statistical Association*, **57**, 327–337.

TAKÁCS, L. (1981). On the 'Problème des Ménages'. *Discrete Mathematics*, **36**, 289–297.

TAKÁCS, L. (1989). Ballots, queues and random graphs. *Journal of Applied Probability*, **26**, 103–112.

TAKÁCS, L. (1992). Random walk processes and their applications in order statistics. *The Annals of Applied Probability*, **2**, 435–459.

TODHUNTER, I. (1865). *A History of the Mathematical Theory of Probability*. Reprinted 1965 by Chelsea, New York.

WALD, A. (1947). *Sequential Analysis*. Wiley, New York.

WHITWORTH, W. A. (1901). *Choice and Chance* (5th ed). Reprinted 1959 by Hafner, New York.

WILF, H. S. (1989). The Editor's Corner: The White Screen Problem. *The American Mathematical Monthly*, **96**, 704–707.

WILF, H. S. (1990). *Generatingfunctionology*. Academic Press, San Diego.

Index